The **HOPE**, **HYPE**, & REALITY of Genetic Engineering

The HOPE, HYPE, & REALITY of Genetic Engineering

Remarkable Stories from Agriculture, Industry, Medicine, and the Environment

JOHN C. AVISE

2004

OXFORD
UNIVERSITY PRESS

Oxford New York
Auckland Bangkok Buenos Aires Cape Town Chennai
Dar es Salaam Delhi Hong Kong Istanbul Karachi Kolkata
Kuala Lumpur Madrid Melbourne Mexico City Mumbai Nairobi
São Paulo Shanghai Taipei Tokyo Toronto

Published by Oxford University Press, Inc.,
198 Madison Avenue, New York, New York 10016

www.oup.com

Library of Congress Cataloging-in-Publication Data
Avise, John C.
The hope, hype, and reality of genetic engineering : remarkable stories from agriculture,
industry, medicine, and the environment / John C. Avise.
p. cm.
Includes bibliographical references and index.
ISBN 0-19-516950-6
1. Genetic engineering—Popular works. I. Title.

QH442.A98 2003
660.6'5—dc21 2003049845

9 8 7 6 5 4 3 2 1

Printed in the United States of America
on acid-free paper

Acknowledgments

Across my 30-year tenure at the University of Georgia, I have felt at liberty to pursue any line of scholarship, however controversial. The science of genetic engineering is a potential minefield for objective inquiry, due to its many social and economic ramifications, and I am deeply appreciative of having been able to examine this field within a climate of complete academic freedom.

The Pew Foundation has supported my work in recent years, and I am grateful to the Pew Fellows and staff for encouragement and inspiration. All or portions of the manuscript were critically read by Joan Avise, Betty Jean Craige, Lee Ehrman, Bob Ivarie, Kirk Jensen, Rich Meagher, Bob Peterson, Jan Westpheling, and several anonymous reviewers, and I thank them all for helpful suggestions. Special credit goes to DeEtte Walker for assistance in all phases of this work. I also want to acknowledge my many students and colleagues, who help keep the intellectual flames alive, and my family, for understanding.

Preface

A scientific revolution is in progress that promises to alter all prior rules on how humankind interacts with the biotic world. Through the laboratory wizardry of genetic engineering, researchers have gained the ability to identify and characterize genes for nearly any biological function, modify these genes and insert them into living cells, swap genetic material freely among species, and even generate perfect genetic copies (clones) of whole plants and animals. By routinely engineering the genetic blueprints of living organisms, scientists today have capabilities far exceeding those of either Dr. Frankenstein or Dr. Doolittle of earlier fiction. Might some real-life GMOs (genetically modified organisms) of the modern era become uncontrollable monsters that turn against their makers, as did the stiff-legged brute of Dr. Frankenstein's creation? Or will they remain our obedient companions, faithfully carrying out their human-assigned roles, as did the pleasant and talkative animal friends of Dr. Doolittle?

This book provides scores of examples of how GMOs with novel and often bizarre genetic capabilities are being generated in the laboratory and sometimes unleashed into the wild. Through case-by-case descriptions of these creatures and the research programs that created them, I hope to inform a wide audience about the activities of modern-day genetic engineers and about how their biological sorceries can affect human lives and the ecology of the planet. My intended readership includes professional biologists, but even more so an interested public and university students in the humanities as well as the sciences. In particular, I hope this book will help stimulate critical thought and discussion in relevant college courses addressing some of the most consequential societal issues of our times.

Research articles on genetic engineering abound in scientific journals and industry documents, but these can be technically difficult. Popular accounts in the news media usually are oversimplified or sensationalized. Here I seek an intelligent yet entertaining middle ground. Using simple and evocative language, but without sacrificing scientific rigor, I hope to bring genetically modified microbes, plants, and animals to life and fairly articulate the hopes and fears they raise in us all. This primer on genetic engineering assumes of the reader only an elementary knowledge of cellular biology and genetics—adequate background information is provided in chapter 2, within the relevant case studies in subsequent chapters, and in a glossary. Also included is an appendix with more extended descriptions of laboratory techniques commonly employed in genetic engineering. Readers interested in methodological details might wish to read this appendix before proceeding to the primary essays in earlier chapters.

My primary goals in this work are to inform, delight, provoke, and intellectually engage the reader in the amazing alchemies of new-age genetic manipulation. The book is organized into a collection of short compositions, each highlighting an ongoing or contemplated effort by genetic engineers to reshape life. In vignettes centered on five major topical areas—microbes, crops, barnyard animals, nondomestic organisms, and humans—I show how genetic engineers are occupied in widely diverse activities such as altering the hereditary makeup of bacteria to produce proteins of medicinal and industrial value; developing crops with genetic resistance to herbivorous insects and chemical herbicides; making perfect genetic copies of prized farm animals; sterilizing unwanted animal pests; contriving plants with a genetic ability to decontaminate toxic waste sites; coaxing embryonic stem cells to rejuvenate worn-out adult tissues; and even cloning human beings.

Emotional responses to the production of GMOs run the full gamut. Even as many biotechnologists, economists, and industry leaders express great optimism for a better planet through genetic engineering, fear reigns in other circles about possible risks to human health, food supplies, and the environment. Closer to the middle of the spectrum, reflective hopes and concerns are expressed by professional biologists as well as by laypeople with a wide assortment of religious and philosophical outlooks.

Thus, another of my goals is to attempt case-by-case appraisals of genetic engineering from several different perspectives. By nearly any standard, the potential benefits to be derived from genetic engineering range from the trivial to the momentous, as do the hazards. Furthermore, the risks and rewards do not always vary in parallel. Some bioengineering agendas are small gambles with huge potential payoffs for society, others are just the reverse, and most fall somewhere in between. To condemn or praise genetic engineering as a single monolithic enterprise is inadequate.

Even when the ultimate goal of a genetic engineering project is beyond reproach, financial gain and/or scientific prestige are the usual proximate forces that drive the enterprise. Unfortunately, money and personal egos can make uncomfortable bedfellows for otherwise well-intentioned science, sometimes biasing the contents even of leading scientific journals. Thus, especially in lucrative high-stake arenas such as biotechnology, all "scientific opinions" should be accompanied by forthright disclosure statements. Here's mine:

Apart from owning a few miscellaneous stock shares, I have no vested financial interests either in, or in opposition to, the biotechnology industry. I have no close social or professional connections to advocates or opponents of genetic engineering, nor within that discipline do I enjoy a high scientific standing to protect. Instead, I hold an academic appointment in the field of evolutionary genetics. The research conducted in my university laboratory, across three decades, has employed many of the molecular tools of genetic engineering, but for quite different purposes: to describe ecological and evolutionary processes, rather than to alter life. I have no personal desire to modify genes, nor any blanket ethical objection to the efforts of those who strive to do so.

Thus, I am a scientific outsider peering into the genetic engineering industry, but I hope to turn this naiveté to an advantage. I hope to bring to the table a genetically informed but relatively detached and objective perspective on the promises and pitfalls of GMOs. My motivations have been a desire to learn more about the topic and a wish to share this knowledge openly with a broad audience. The original Dr. Frankenstein worked in secret, hiding from the torch-carrying mobs who were suspicious of the late-night activities in his laboratory. This book will provide a more civilized way to illuminate the labors and the biological inventions of today's real-life genetic engineers.

Contents

The **HOPE**, **HYPE**, & **REALITY** of Genetic Engineering

1

A Tale of Good and a Tale of Evil

This book is about humankind's recent attempts, using recombinant DNA technologies, to modify the genetic makeup of living organisms. It describes how scientists have learned to isolate and transfer DNA from one species to another, generate clonal (genetically identical) copies of individual genes and sometimes whole organisms, and otherwise exercise godlike influences over the hereditary material of microbes, plants, and animals, including humans.

Using a case-history format, this book traces the rise of genetic engineering during the latter part of the twentieth century, addressing both the good and the bad of dozens of genetic engineering projects initiated to date. This empirical track record reveals many of the triumphs and disasters of genetic engineering, its amazing feats and some blazing defeats, and the field's beauty marks and its blemishes both small and large. The two stories in this opening chapter—one the happy narrative of a successful little genetic engineering venture, the other a chronicle of sheer madness—serve to introduce the book's style and format, as well as to illustrate the wide range of potential societal outcomes when humans attempt to seize control of life's genetic reins.

A New Papaya

The papaya tree (*Carica papaya*) is native to the lowlands of Central America, but it is now grown commercially at scattered sites throughout the world's tropics and subtropics. The tree and its fleshy fruit are natural hosts to a pathogen, the papaya ringspot virus (PRSV), which can seriously reduce fruit production in papaya orchards. Different PRSV strains take a heavy toll on papaya production in many tropical countries, from Jamaica, Venezuela, and Colombia to Taiwan, Thailand, India, and Queensland, Australia. Plant-infecting aphids spread PRSV, and sometimes the infestations are so severe that the only traditional recourse has been to abandon old orchards and plant new ones in uninfested areas. In Brazil, for example, over time the papaya industry has shifted from one geographic area to another to present moving targets for this virus.

A similar situation occurred in the Hawaiian Islands, where papaya is the archipelago's second largest fruit crop, generating revenues of more than $45 million annually. The virus was discovered on the islands in the 1940s, and during the 1950s it virtually eliminated papaya production on Oahu. In response, much of the industry moved to the virus-free Puna district on the Big Island, where the crop thrived for the next several decades. However, in May 1992, a long-feared viral invasion of the Puna region finally occurred. From 1993 to 1997, PRSV economically destroyed plantation after plantation, cutting annual papaya production in Hawaii by nearly 40%.

Scientists at Cornell University then identified, isolated, and cloned a gene (encoding a cell-surface or "coat" protein) from the papaya virus and inserted it into cells of experimental papaya trees. In its new plant home, the viral gene continued to produce coat protein, and this stimulated the genetically modified (GM) papaya trees to develop strong genetic resistance to the papaya virus. This foreign protein acted much like a vaccine, and the trees were in effect immunized against PRSV. The GM papayas were released for commercial use in May 1998, and, in 1999, Hawaii saw its first increase in papaya production since the arrival of the ringspot virus at Puna nearly a decade before.

The successful engineering of Hawaiian papaya is a mostly happy tale with no evident downsides, so far. Long before the new GM strains (SunUP and Rainbow) were generated, people inadvertently ingested the coat protein from virus-infected papayas without ill effect, so the GM version of the fruit should carry no untoward consequences for human health. Furthermore, it seems unlikely that the coat-protein gene will move naturally to unintended sites because the papaya tree is pollinated primarily by insects (not wind). Thus, the gene-carrying pollen (male gametes or sex cells) seldom move far, and, in any event, *C. papaya* hybridizes only rarely with other *Carica* species. There is one concern, however: The virus might evolve countermeasures to the genetic resistance of GM trees, in which case the crop could come under renewed attack.

In the case of the GM papaya, the progression from pure science to practical application was also laudatory. The scientists involved were not associated with a commercial seed company, so economics did not unduly dictate the goals or approach of the basic research program. Regulatory agencies (the U.S. Department of Agriculture, Environmental Protection Agency, and the Food and Drug Administration) and other constituencies (including the University of Hawaii, Cornell Research Foundation, and a Hawaiian papaya growers' association) all worked together to effect the transfer of this GM technology from the laboratory to the field. Finally, the intended beneficiaries of the endeavor—papaya consumers and Hawaiian farmers—mostly have welcomed these genetic engineering achievements.

Microbiological Terrorism

During World War II and the next two decades, in the age before genetic engineering, scientists at the U.S. Army Medical Research Institute of Infectious Diseases (USAMRIID) at Fort Detrick, Maryland, clandestinely researched biological agents that in principle could far surpass the killing power of the most potent traditional weapons. So too did scientists in the former Soviet Union and at least a dozen other countries. No program was larger than that of the USSR's Biopreparat Agency, where microbes for fearsome diseases such as smallpox, plague, anthrax, tularemia, Q fever, brucellosis (undulant fever), Marburg virus, and Lassa virus were weaponized into bombs and other delivery vehicles. According to one former Soviet insider, the Biopreparat had hundreds of tons of anthrax ready for use at any time, and a 20-ton stockpile of the smallpox virus. It also reportedly tested such microbes in the open air, for example, on monkeys at Resurrection Island in the Aral Sea.

Biological warfare and terrorism, defined as the use of disease agents for hostile purposes, were abjured by the United States in 1969, during the Nixon administration, and formally outlawed by the International Biological Weapons Convention Treaty of 1975. In 1992, Boris Yeltsin finally shut down weapons research at the Biopreparat, but no one is entirely certain what became of all the facility's deadly bacteria and viruses. And, of course, no treaties or directives can ensure that all such biological weapons research has ended. Enforcement and verification are far from perfect even in openly democratic nations, much less in dictatorships and terrorist organizations.

Here's a chilling scenario describing the harm that could be wrought by even a single nasty disease strain. Suppose that a rather benign bacterium, producing only mild intestinal disorders, somehow loses the genes that encode adhesins, proteins that bind microbes to our intestinal walls. The GM strain thereby is freed to move throughout the body. With deadly speed, it invades the liver, spleen, lungs, and other organs, overwhelming the body's immune defenses. People die agonizing deaths as their lymph nodes balloon into ugly masses, or their lungs fill with fluids. The deadly bacteria, spread far and wide by alternative hosts like fleas and rats, or by casual human contact, soon find their way to such places as Europe, where they sicken 50% of the population and kill 1 person in every 4, to Asia where they kill 10 million more, and so on. Eventually, around the world, hundreds of millions of people perish.

Although this nightmarish disaster might seem like science fiction, it is paralleled closely by historical fact. The real ancestral microbe was *Yersinia pseudotuberculosis*. The genetic changes it underwent occurred in nature, probably in the past 1500 to 20,000 years. During that time, the bacterium evolved, as described, into a far more deadly form that was given the name *Y. pestis*, the bacterial agent of plague, or black death. Several times during the Middle Ages, and as recently as the twentieth century, the plague reached pandemic scales, wiping out huge segments of humanity.

In response to the threat of such natural biological disasters in modern times, and also in response to fears that *Y. pestis* could be delivered intentionally by bombs or other weapons, scientists at USAMRIID succeeded in developing effective vaccines for the most dangerous pneumonic form of the microbe. Concurrently, however, scientists at Biopreparat were engineering a new strain of bacterial plague with genetic resistance to several antibiotic drugs. In theory, these genetically altered bacteria, if delivered by aerosol attack, could produce more deaths than the most horrific of the natural plagues in human history.

What is to prevent terrorists or others from likewise engineering and unleashing such dreadful disease agents? Unfortunately, in this context, the relevant gene-manipulating techniques are fairly inexpensive and simple. It would be relatively easy for someone to mix and match genes from different sources, using standard recombinant DNA methods, to engineer bacterial strains that could be all but impossible to control with conventional drugs. Indeed, nature routinely conducts such GM experiments, without direct human assistance, and has succeeded on occasion with the emergence of microbial strains displaying joint immunity to multiple antibiotics such as penicillin, tetracycline, erythromycin, methicillin, and even vancomycin (one of the more recent lines of medical defense). The "superbugs" arise step by step when microbial populations gain mutations, and sometimes swap them through various natural means, that happen to confer genetic resistance to the antibiotic drugs.

Another GM approach for biological terrorists or rogue nations would be to construct microbial agents of human disease, such as poliovirus, from simple chemical building blocks. Naturally occurring polioviruses have plagued humans throughout the ages, killing and paralyzing millions of victims until a worldwide vaccination campaign, begun in 1988, almost exterminated the culprit. Could someone with nefarious motives resurrect this or other deadly disease agents in the laboratory? Such nightmares became more plausible in 2002, when scientists (presumably well-intentioned) published an article in the online journal *Science* reporting that they had artificially synthesized a poliovirus from scratch. The poliovirus is exceptionally tiny, its genome (full suite of genetic material) a mere 7741 genetic letters long. Working from a publically available database describing the full genomic sequence of the virus, the scientists had used commercially available machines to chemically synthesize and join together bits and pieces of the viral genome into a functional whole. To everyone, this macabre experiment sent a clear message: The vector of a horrible human disease can be created *ex nihilo* in the laboratory.

As knowledge continues to grow concerning the genetics of microbial disease agents (e.g., the complete 4.6-million-letter genome of *Y. pestis* was revealed in 2001), villainous minds could think of many additional ways to create or genetically alter microbes for gruesome pathogenicity. At the same time, further genetic information about disease-causing microbes should advance well-meaning research into vaccines, antibiotic drugs, rapid-detection methods for pathogenic organisms, and other countermeasures to bioterroristic threats.

For example, a recent report in *Nature Biotechnology* (Maynard et al.) showed how recombinant DNA techniques might be used to generate therapeutic antibiotics for neutralizing toxins from the anthrax bacterium (*Bacillus anthracis*). The development of such public safety measures has been an underlying motive in efforts to sequence the full genomes of disease-causing bacteria such as *Vibrio cholerae* (the agent of cholera), *Brucella suis* (brucellosis), and *Staphylococcus aureus* (enterotoxin B). Such efforts, although laudable, are not without risks. As the field of genetic engineering grows and offers greater opportunities for societal betterment, so too do potentials for abuse of this technology. Some parallels may exist between the current expansion in molecular genetics with that which occurred in nuclear physics beginning several decades ago. If even a few scientific experts direct their powerful technical skills toward abominable ends (e.g., proliferation of nuclear weapons in the case of physics), the damage they do could outweigh the collective societal good done by legions of scientists with commendable motives.

Furthermore, the dangers from genetic engineering arise not only from those with evil intent, but also from simple mistakes or lapses in judgment by scientists with good intentions. An example of the latter recently came to public attention when the British government fined the Imperial College of London for failing to follow research guidelines and compromising public safety in its attempts to speed development of vaccines and drugs against the hepatitis C virus, which infects about 200 million people worldwide. Using recombinant DNA methods, researchers at the college had spliced key genes from dengue fever virus into the hepatitis virus, thereby creating a potentially deadly hybrid type of microbe. Thankfully, nothing bad came of this genetic faux pas.

This wasn't the first incident of its sort. Just a few months earlier, while trying to find a simple way to sterilize rodent pests in Australia, scientists generated great alarm when they "accidentally" engineered a deadly mousepox virus. What was to stop evil-doers from accomplishing the same feat with the human pox virus?

Clearly, genetic engineering, with malice, must be of great concern to societies. If there is a ghoulish saving grace of sorts, especially in the context of bioterrorism, it is that many natural (nonengineered) agents of human disease already are hideous and proliferative enough to imperil massive numbers of people: witness smallpox, and the AIDS virus. Thus, if an evil government or terrorist group wished to create havoc by dispersing microbial diseases (of people, or of our agricultural crops and animals), it might find no compelling need for genetic engineering per se. More simply, it could explore physical means to mass deliver any of the numerous gruesome microbes already found in nature.

In efforts to thwart one such possibility, the Bush administration in 2002 rush-ordered the manufacture of sufficient smallpox vaccine to help protect significant numbers of U.S. citizens from a potential bioterrorist attack. This brings back haunting memories of earlier times when, in effect, biological warfare via the smallpox virus actually was waged against North Americans. When Hernando Cortes conquered Mexico in the early 1500s, the smallpox virus that his men inadvertently introduced to the continent killed far more Native Americans than did all European guns and swords. Later, during the American Revolutionary War, another smallpox outbreak killed about 130,000 colonists. The epidemic was so horrific that George Washington suspected (incorrectly) that the British army had deliberately spread the infection. Clearly, biological agents are fully capable of mass destruction.

2

Framework of an Unfolding Revolution

Genetically modified organisms (GMOs) are plants, animals, or microbes whose genes have been deliberately and directly altered by humans. A given genetic remodeling job may be as focused as swapping one gene for another or as extensive as generating a perfect replica of an organism's genome (the full complement of DNA within any of its cells). The biological consequences of genetic engineering may be as subtle as slightly amending the metabolism of an individual animal or plant or as dramatic as recasting the evolutionary trajectory of an entire species or even the ecosystem of which it is a part.

As used in this book, "genetic engineering" is not entirely synonymous with "biotechnology." The latter refers to any technological applications for living entities, whether genetically modified by humans or not. Thus, the use of naturally occurring yeasts to produce alcoholic beverages or cheeses is a biotechnological application, but not an example of genetic engineering. The biotechnology industry routinely uses bacteria to biomanufacture a wide range of pharmaceuticals and other marketable products, but only some of those microbes have had their DNA intentionally manipulated by geneticists.

Depending on one's outlook, the technical wherewithal to engineer life, purposefully and sentiently, is either an angel's blessing or devil's curse. It is a capability borne of the revolution in recombinant DNA technology over the last three decades. Genetic engineers now routinely alter genes or the cells that house them through a variety of biochemical and physical methods, typically using molecular genetic tools that nature provides (see appendix). Viruses and bacteria are unwitting accomplices in many of these genetic manipulations. Intended recipients of the newly engineered genes may be microbial organisms, plants, or animals, including humans.

The Purview of Genetic Engineering

Conscious genetic engineering differs from incidental genetic modification, such as may occur when people introduce mutagens (like radioactive wastes or asbestos) into the environment. Such mutation-inducing agents certainly can alter genes, but the genetic changes normally are haphazard, unintended, and damaging. Purposeful genetic engineering also differs from inadvertent genetic alterations that may happen in natural populations when human activities alter ecological selection pressures. Thus, bacterial strains often evolve a genetic resistance to widely used antibiotics such as penicillin, but this would not qualify as a response to deliberate genetic engineering. Nor would the evolution of insecticide tolerance by a mosquito population exposed to the chemical dichlorodiphenyltrichloroethane (DDT).

Modern genetic engineering should also be distinguished from purposive bioengineering first practiced about 10,000 years ago, when our ancestors began domesticating plants and animals. Via selective breeding over the centuries, people genetically altered numerous plant species to yield finer food and fiber, altered livestock to produce better meat and milk, and altered dogs and cats to improve their desirability as companion pets. Such artificial selection (a human-mediated analogue of Darwinian natural selection) is a slow and indirect method compared to the fast and focused recombinant DNA technologies ("gene-splicing" methods) of today. Furthermore, artificial selection acts only on the available stores of genetic variation in closely related creatures that can interbreed, whereas modern genetic engineering can in principle swap genes freely among any living creatures, from microbes to trees and mammals.

As described in the appendix, bacteria and viruses long ago mastered the art of cutting and pasting genes from one biological source to another, often instilling the recipient cells with new functional capabilities. The molecular tools and workshops that microbes evolved for manipulating DNA are precisely those that humans now co-opt to serve our own genetic engineering missions. When biotechnologists argue for the social acceptance of industry methods, they often note that genetic engineering in research laboratories differs little, fundamentally, from that long practiced by nature. Biotechnologists contend that the primary distinction is that genetic engineering by nature is mindless and unconscious, whereas that by humans is mindful and well intentioned.

The stated or implied goals of most genetic engineering programs are laudable: to increase the quantity or quality of food crops, to clean up environmental toxins and wastes, or to improve human health. If a GM project should happen to go awry and harm people or the environment, inadequate foresight or poor scientific expertise normally would be to blame, rather than ill intent. This would be of small consolation to injured parties, however. Furthermore, a few genetic engineers might have shameful motives, ranging from unbridled greed, to distasteful eugenic missions, to the creation of biological weapons.

On the other hand, genetic engineering offers many unprecedented opportunities in medicine, agriculture, and environmental stewardship. Thus, to be blind to the possibilities of GMOs would be irresponsible. People can pontificate at length about the merits and demerits of genetic engineering on philosophical grounds, but the fact remains that the genetic genie is well out of the bottle.

Preliminary Genetic Background

The relative ease with which today's genetic engineers can manipulate and swap genes among species is made possible by the near universality of DNA (deoxyribonucleic acid), the genetic material of life on earth. In 1953, James Watson and Francis Crick (following key discoveries by Maurice Wilkins and Rosalind Franklin) deduced this molecule's simple yet elegant structure: a graceful double-helix with two intertwined threads each composed of long ordered strings of the nucleotides adenine (A), guanine (G), thymine (T), and cytosine (C), each A in one strand paired with a T in the other, and each G paired with a C. [In some viruses, RNA (ribonucleic acid) is the genetic material instead, but RNA is structurally similar to DNA, one difference being that it usually exists as a single thread rather than two.]

The human genome is composed of about 3 billion nucleotide pairs. This is a fairly standard count for vertebrate animals. Near the other end of the continuum, the genome of the common bacterium in our gut, *Escherichia coli*, is only 4.6 million nucleotides long. Each genome in existence today, regardless of its size or biological source, is merely the latest link in an unbroken chain of evolutionary descent with modification, the current tip on one of nature's continuous branches of ancestry extending back across 4 billion years to when life on earth first appeared. Each genome is also a working blueprint prescribing the construction and metabolic operations of a living individual.

Nucleic acids (DNA and RNA) are nature's heritable manuals, storing the information required for life and passing that accumulated evolutionary knowledge across the generations. Proteins, the functional and structural workhorses of cells, are primarily what these manuals prescribe. Thousands of different kinds of proteins are at work in each cell. Many are enzymes, organic catalysts that grease nearly all molecular reactions, keeping a cell's biochemical pathways from freezing up, much like motor oil lubricates moving parts in a car's engine. Other proteins contribute to a cell's physical fabric, such as its walls, membranes, and molecular transportation capabilities.

Each protein is made up of one or more polypeptide chains, and each polypeptide is composed of a long string of amino acids, the order of which is specified by the coding regions (exons) of a particular gene. Amino acids come in more than 20 types, each ultimately dictated by a different triplet of nucleotides in the cell's DNA (for example, CCT specifies glycine and CGG specifies alanine). Thus, a polypeptide's amino acid sequence is colinear with and stipulated by the linear nucleotide sequence in the gene that encodes it. Transcription and translation are the sequential cellular processes by which the precise specifications in the nucleotide sequences constituting genes are converted to corresponding amino acid sequences in the resulting proteins.

During transcription, messenger RNA (mRNA) molecules are composed from the DNA text. Each mRNA is like a miniature transponder, receiving and then faithfully transmitting a coded signal emanating from a segment of DNA. During translation, that molecular message is converted to a polypeptide chain. In eukaryotic species (all plants, animals, and some microbes), transcription takes place within a membrane-bound nucleus, the cell's command center. Translation occurs on special structures known as ribosomes present in the cytoplasm, the part of a cell outside the nuclear membrane. In prokaryotic creatures (including all bacteria, whose cells lack a well-defined nucleus), transcription and translation are less separated within the cell, but are otherwise similar.

In nature's nucleic acid alphabets, nucleotides are the four kinds of letters (G, A, T, and C) that, variously ordered, compose every genetic word, sentence, paragraph, and chapter in each organism's operations manual. In the genetic engineering trade, scientists edit these tomes, substituting a biochemical letter here, inserting a word or sentence there, transposing paragraphs, duplicating particular passages or sometimes the whole text, and, in general, attempting to tweak or revamp particular instructions in the code. Natural evolutionary processes (most notably mutation and selection) have forged all life as we know it. Today's genetic engineers aim to improve the outcomes, typically according to the standard of how well a particular GMO satisfies human needs.

In the early 1970s, geneticists discovered how to isolate pieces of DNA from any species, splice the molecules together in test tubes, and reinsert the recombinant genes into living individuals. Since then, laboratory approaches to genetic engineering have become highly diverse, but all can be categorized as attempts to modify genetic material already present within an organism or to introduce exogenous genes from foreign biological sources. Often, the inserted DNA is derived from another species. The recipient in any genetic "transformation" procedure is referred to as a transgenic (or GM) organism, and the inserted DNA is called a transgene. Today, transgenic creatures are generated routinely in government laboratories, public and private companies, and universities. Some of the procedures are so inexpensive and technically simple that, for better or for worse, they can be conducted in garages or in high-school science classrooms.

Broadly speaking, genetic engineering is nothing new. Nature has been perfecting many of the fundamental techniques for hundreds of millions of years; as, for example, when bacteria use special enzymes to chop up invasive viral DNA, or when a virus inserts itself into an animal or plant cell and makes copies of itself by co-opting the host cell's replicative machinery. Indeed, most of the laboratory procedures used by genetic engineers were either borrowed directly or conceptually prompted by the DNA-manipulating methods that microbes evolved long ago (see appendix).

In many cases, producing a transgenic organism is the technologically easy part. Often the greater challenge is to engineer the transgene so that it functions as intended in its new biological house. Accordingly, much of the research effort put into genetic engineering is directed toward designing and then monitoring transgenes to see that they function properly at several levels: inside the cell and its metabolic pathways; within the developmental milieu of the organism; and in the context of the species' ecological setting. Therein lie some of the greatest challenges for the genetic engineering enterprise.

Throughout the lifetime of any organism, the products of multitudinous genes (approximately 40,000 in the human genome) normally cooperate in elaborate molecular dances whose choreographies have been honed by the cumulative actions of natural selection across eons of evolutionary time. Little wonder that genetic alterations (be they naturally occurring mutations or artificially introduced transgenes) are far more likely to disrupt than to refine these exquisite molecular ballets. However, nature's pathways are also littered with failures. Genomes, far from being perfect constructs created by an omnipotent designer, are products of mechanistic evolutionary processes that are entirely devoid of foresight, intelligence, and reflective concern. It is no surprise that so many early embryos (in humans and other species) naturally abort from genetic disorders, that serious genetic disabilities afflict so many of those who escape this initial slaughter, that senescence and death of the individual are phenomena universal to life, or that the fate of most (and eventually all) species is extinction.

It may be audacious of humans to try to improve life's operations through genetic modification. However, hubris is hardly foreign to our species. For more than 200,000 years, people have sought and gained considerable dominion over nature, modifying natural environments, in effect, to suit our genes. Now, suddenly, we find ourselves with the technological capacity to modify genes to better meet our physical and cultural needs. Will genetic engineering revolutionize our existence? In producing GMOs, will we exercise restraint when dangers loom, yet still summon the courage to pursue what may be glorious opportunities? Indeed, will we have the wisdom and foresight to know the difference?

The Boonmeter

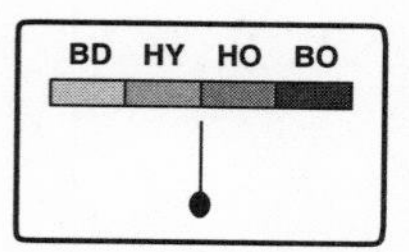

Chapters 3–7 contain more than 60 essays, each describing an ongoing or contemplated genetic engineering endeavor. At the end of each essay, I include an appraisal gauge or "boonmeter" that summarizes my overall impression of the current societal merit or status of that particular project. The meter's needle can point anywhere along a continuum of categories:

(a) *boondoggle* (*BD*): a bad idea to begin with, or, sometimes, a decent concept but an abject failure in practice (to date);
(b) *hyperbole* (*HY*): promoted with considerable fanfare, but yet to live up to all the hoopla and unlikely to be of great benefit to broader society, unless new breakthroughs or shifts in orientation occur;
(c) *hope* (*HO*): an approach that holds considerable promise, but for one reason or another has not yet come to extensive fruition; or
(d) *boon* (*BO*): a research domain that already has produced tangible benefits for society, and with few or no evident downsides.

This boonmeter has several provisos: First, each judgment is my own personal (and often debatable) summary opinion, meant only as a possible starting point to stimulate discussion. For example, human reproductive cloning may be a questionable idea overall, yet nonetheless may have merit in special instances, such as when it enables infertile couples to have children. Second, each appraisal applies only to the state of affairs at the time of this writing, so with new developments or convincing arguments to the contrary, my provisional evaluations could change dramatically (indeed, I hope they will do so for research areas that I have rated poorly for now). For example, engineering the human germline (reproductive cells) has been hyperbole to date, but in the future might conceivably become a powerfully beneficial way to circumvent horrible genetic diseases.

Third, I try in the boonmeter to summarize the outcome from a broad societal perspective, more so than from the immediate vantage of close stakeholders. For example, several recombinant DNA products in agriculture have created windfall profits for agribusinesses, but their lasting benefits to farmers, consumers, or the environment are dubious in some cases. Finally, some subjects are especially hard to categorize because, for example, they are highly touted and promising, but they also carry significant risks, and the jury is still out on their final merit. On the appraisal gauges, I have often depicted these as neutral for now (i.e., arrow pointing between hype and hope).

3

Engineering Microbes

Microbes have supplied more than just the molecular tools and the conceptual inspiration for genetic engineering (appendix), they also have been harnessed to manufacture some of the most lucrative commercial products that have emerged to date from the GM industry. This chapter provides several examples.

First, however, the usual meaning of gene "cloning" in the context of microbial genetic engineering should be made clear. In microbes, gene cloning (the production of "carbon copies" of a particular segment of DNA) typically is accomplished when genetic engineers insert a foreign gene into plasmids, which are tiny circular pieces of DNA that reside within bacterial cells. When the GM bacteria then do what comes naturally (divide and multiply asexually), they automatically clone the transgene as well as their own native DNA. As we shall see, genetic engineers often use such GM bacteria and their purposefully cloned transgenes to produce commercial quantities of various proteins of pharmaceutical or industrial importance.

Insulin Factories

Insulin, the diabetes-treating hormone, is now mass-manufactured from transgenic bacteria that have been engineered to carry and express the human insulin gene. The human insulin saga is of historical interest because it provides one of the first success stories in genetic engineering and also illustrates the broader trials and tribulations of the neophyte GM enterprise.

In the spring of 1976, in Indianapolis, the pharmaceutical giant Eli Lilly convened a national scientific symposium about insulin. For decades, Lilly had purified this pancreatic protein, for human therapeutic purposes, from slaughtered cattle and pigs. When administered by injection, animal insulin enables diabetics to metabolize sugars that otherwise accumulate in their bodies at debilitating and sometimes fatal levels. Lilly had done well in the insulin trade, tallying about $160 million in annual sales to a market of more than 1 million insulin-dependent diabetic sufferers in the United States alone. But Lilly's business charts identified an ominous trend: projected insulin demand someday might outstrip the available supply from farm-animal tissues. Furthermore, as a therapeutic drug, animal insulin was less than ideal because its structure differed somewhat from that of human insulin. This could cause problems, as, for example, when diabetic patients developed antibodies to animal insulin.

The Indianapolis meeting had a stellar lineup of insulin experts as well as leaders in the infant field of molecular biology. The latter included Howard Goodman, known for his pioneering role in the characterization of bacterial restriction enzymes; William Rutter, an expert on molecular functioning in the pancreas; Argiris Efstratiadis, soon to be famous for inventing methods for isolating genes and for synthesizing gene segments *in vitro* (outside living bodies); and Walter Gilbert, who four years later would receive a Nobel Prize in chemistry for contributions to the experimental manipulation of DNA. This impressive scientific gathering provided an early intellectual spark for genetic engineering research.

Participants came away from the meeting fired by a realization that it might be feasible to isolate the insulin gene from the tangled jumble of DNA in mammalian cells, clone that gene by splicing it into bacteria, and perhaps even coax the GM bacteria to produce human insulin in commercial quantities. Soon, a scientific race began as researchers got caught up in an insulin sweepstakes that promised huge scientific as well as financial rewards to the winners. Major participants in this race were a Gilbert–Efstratiadis team at the University of California at San Francisco, a Goodman–Rutter team at Harvard, Herb Boyer (another noted molecular biologist in California, who a month earlier had co-founded Genentech, one of the world's first genetic engineering companies), and industry scientists at Genentech and Lilly, for example.

Within a year, the University of California group announced a major breakthrough. Using pancreatic cells from a special strain of rats, they had managed to isolate the protein-coding DNA sequence for rodent insulin and insert and amplify it in a bacterium. In other words, they had cloned the rat insulin gene, much to everyone's amazement. About one year later, the Harvard group did them one better by contriving a bacterial strain that expressed (turned on) the rat insulin transgene. In essence, the GM microbes produced a protein of potential use in the treatment of diabetes.

Technically, however, what those GM bacteria actually produced was proinsulin, an insulin precursor molecule. At about that same time, the California and Genentech teams were racing to find ways to align, snip, and girder such precursor molecules so as to get bacteria to make bona fide mammalian insulin. Meanwhile, Genentech and the University of California made independent agreements with Lilly regarding how the technologies might be applied to clone and express the human insulin gene for commercial markets. Finally, in 1982, Lilly began to market genuine human insulin produced by GM methods. Sold under the name Humulin, this was the first biopharmaceutical product of medical value to be engineered by recombinant DNA procedures. Today, according to Lilly, more than 99% of all insulin used in the United States comes from genetically engineered sources. Total U.S. sales of this drug amount to more than $3 billion annually.

It would be naive to think that this entire research effort was cordial and collaborative. To the contrary, it was often highly competitive and even cutthroat, as recently became apparent during lurid courtroom disclosures in a major lawsuit. For the groundbreaking technologies that its faculty devised during the course of cloning the insulin gene, the University of California won two U.S. patents that Lilly ignored as the company proceeded with marketing human insulin from recombinant bacteria. To tap into some of the windfall profits, the University of California sued Lilly in 1990. This led to an eight-year court battle with squadrons of attorneys representing both sides and charges of misconduct flying back and forth over technical scientific and legislative matters. In the end, a federal judge ruled in favor of Lilly, yet also exonerated all of the participating researchers of any scientific misconduct.

Beyond stiff competition per se, the 1970s were trying times for the genetic pioneers for another reason. Even as researchers began conducting their seminal GM experiments, American society had begun debates over the fundamental merit of any and all genetic engineering. For example, in Cambridge where the Harvard team worked, the city council held heated public hearings that in 1976 precipitated a critical seven-month moratorium on certain kinds of recombinant DNA experiments.

On the national scene, scientists already had debated such matters for several years, one result being a series of research guidelines by the National Institutes of Health (NIH) concerning recombinant DNA procedures. This proved crucial in the insulin case because in 1976 the NIH had not yet certified as safe one of the plasmid cloning vectors (see appendix) that was central to the ongoing insulin research, especially in California. Ambiguities related to this certification process surfaced again in the aforementioned court battle between the University of California and Lilly. Thus, especially in those early years of genetic engineering (but continuing to a lesser extent today), the sociopolitical climate in the United States was a daunting hurdle for basic and applied research on recombinant DNA.

Notwithstanding these several controversies that accompanied development and marketing of GM insulin, the enterprise overall merits recognition

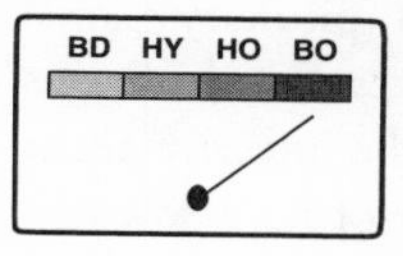

as an unequivocal boon, both in terms of producing a valuable medical product and in having pioneered ideas and research approaches that were critical to the broader development of the genetic engineering field.

A Growth Industry

Insulin was the first human hormone to be produced in commercial quantities by transgenic microbes. The second, following close on its heels, was human growth hormone (hGH), also known as somatotropin, derived from Greek words for "body" and "nourishment." Like insulin, therapeutic hGH has a fascinating and somewhat troubled history.

In the human body, hGH normally is produced by the pituitary gland (a small cherry-shaped body attached by a stalk to the base of the brain), and then passed into the bloodstream, where it exerts influence over nearly all tissues and organs. Notable among these effects are the stimulation of natural bone and muscle growth. A severe deficiency of hGH stunts growth in children (producing one of many forms of dwarfism), but also can produce symptoms of extreme fatigue, anxiety, depression, and malaise in anyone. It is estimated that about 3 in every 10,000 adults are seriously deficient for hGH, and that approximately two-thirds of those shortages began in childhood. The usual culprit is underactivity of the pituitary gland (hypopituitarism) such as may result from a brain tumor, surgery, radiation therapy, or problems with blood supplies to that organ.

In the 1920s, growth hormones (GHs) were purified from mammals and were shown to be effective in increasing growth in rats and dogs. In the 1930s, first attempts to treat human dwarfism using GH from bovine pituitary glands failed. Not until the 1940s was it fully appreciated that primates, humans included, respond only to primate GHs. In 1958, the first clinical use of human GH to treat hypopituitary dwarfism took place. During a 10-month trial, a 17-year-old patient administered the drug grew much faster. A rush to administer hGH to undersized children followed during the 1960s and 1970s, and thousands were injected with the compound.

Although often effective in stimulating growth and allowing children with hypopituitarism to reach normal stature, this approach had serious problems. First was the matter of availability. At that time, pituitary glands from deceased donors were the only source of hGH, but human cadavers were in limited supply and individually yielded only tiny quantities of the drug. Thus, the expense of hGH treatment was astronomical. A second problem followed directly from the first. In the biochemical isolation of hGH, pituitary glands from thousands of cadavers first were pooled into large batches. But the brains of some dead donors later were discovered to have carried the infective agent for Creutzfeldt-Jakob disease (CJD), the human equivalent of mad-cow disease. Furthermore, that agent had not been fully removed during the hGH purification process. The hGH distributed in the United States by the National Hormone and Pituitary Program, for example, was derived from a total of 1.4 million pituitary glands, an estimated 140 of which might have been infected. One result was the death from CJD of several treated children and lifelong fear in thousands of families that their children might have contracted this slowly developing neurodegenerative illness.

Research on recombinant DNA human growth hormone (rhGH) began in the late 1970s as a way to circumvent the problems of expense and disease contamination that had plagued conventional hGH therapies using cadaver-based material. Central participants included several of the same research scientists and companies also involved in the development of human insulin. Howard Goodman helped develop methods that resulted in the successful cloning of the rhGH gene in bacteria in 1979. In 1985, Genentech began to market rhGH for children with pituitary deficiencies, as did Eli Lilly in 1987. The rhGH story also has its own lawsuit drama. In 1999, the University of California won a $200 million settlement from Genentech for patent infringement related to Protropin, the company's own costly rhGH drug.

Microbe-produced rhGH drugs now on the market, including orally administered forms, are free of CJD, essentially unlimited in supply, and come at a price that more people can afford. They have helped many children with hypopituitarism grow up to lead more normal lives. In 1996, the Food and

Drug Administration (FDA) also approved the drugs for adults with brain tumors or other factors causing pituitary disease and hGH deficiency.

Notwithstanding the medical and commercial success of rhGH, some troubling ethical questions have arisen. First, should societies permit rhGH use by children of short stature but without serious hypopituitarism? Many parents seek growth-enhancing drugs in the belief that their children thereby might gain higher social standing or economic opportunities in life (lifetime income is correlated with physical stature, some studies suggest). Second, since rhGH helps build muscle mass, should it be used by athletes or others to enhance physical performance? At the 1998 Olympic Games in Sydney, Australia, for example, and in professional baseball in the United States, the wide use of somatotropinlike steroids was suspected and has been verified. Finally, what are the long-term health risks and benefits of high supplemental dosages of GH? Because the market drugs have been available to a large audience for less than 20 years, no one fully knows.

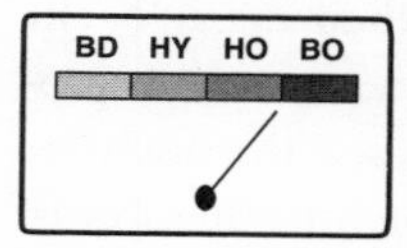

Microbial Factories for Pharmaceutical Drugs

The following thumbnail history of microbiology highlights the fact that our knowledge of microbes and their signal role in human health is remarkably recent. In 1836, the German physiologist Theodor Schwann attributed processes of putrefaction and fermentation to microorganisms, a radical notion that the French biochemist Louis Pasteur experimentally confirmed in the 1860s. In 1882, the German Robert Koch developed a method to isolate pure bacterial cultures from the ill and pinpointed the tuberculosis germ as the first known microbe to cause a human sickness. In 1885, the English surgeon Joseph Lister pioneered the use of microbial disinfectants during surgery, thereby profoundly lowering patient death rates. In that same year, Pasteur formally added vaccination to our medicine bag when he successfully treated a nine-year-old Alsatian boy for rabies. In 1892, the Russian Dmitri Ivanovski discovered viruses, even tinier disease agents. In 1917, phage viruses (which infect bacteria) were discovered by Englishman Frederick Twort and Frenchman Félix D'Herelle. In 1928, another Englishman, Alexander Fleming, isolated penicillin, an antibacterial compound produced by molds. In 1941, Ukrainian-born Selman Waksman coined the word "antibiotic" to describe drugs that kill bacteria.

More recently, microbial organisms also have played key roles in the recombinant DNA revolution. In 1951, the American Joshua Lederberg was among the first to discover that DNA pieces often transfer naturally between bacteria (e.g., via viruses or by the exchange of plasmids, small circular pieces of DNA found inside bacterial calls). Now, viruses and plasmids are used routinely by the biotechnology industry for a variety of gene transfer and DNA cloning procedures (see appendix). Restriction enzymes and ligases, key bacterial enzymes used by genetic engineers to construct recombinant DNA molecules, were discovered in 1967 and 1968, respectively. In the early 1970s, the radical notion that life forms could be licensed was introduced in a U.S. Court of Customs decision allowing a patent for an oil-eating bacterium. This view, upheld in a 1980 decision by the U.S. Supreme Court, opened a commercialization floodgate for GM products. In 1975, the first mammalian gene was introduced to a bacterium for artificial cloning purposes, and, in 1982, human insulin became the first drug to be marketed from transgenic bacteria.

In the last two decades, GM microbes have proved their merit in the pharmaceutical industry. In many cases, the living microbes are used to produce recombinant drugs of medical importance. In other cases, the microbes supply utilitarian enzymes or serve as delivery vectors for introduction of transgenes into plants or animals that then manufacture a useful drug. Worldwide sales of therapeutic drugs made using recombinant DNA techniques has totaled at least $100 billion annually in recent years. Below I describe some of these high-tech products that are already market-approved and the many human disease conditions and ailments these pharmaceuticals are intended to treat.

Recombinant follicle-stimulating hormones (e.g., Follistim) are used to treat human infertility. Biotech human albumin (Albutein) finds application in cardio-bypass surgery. Various monoclonal-antibody products (such as Zenapax and Orthoclone OKT3) help ameliorate kidney rejection in surgical transplants. Several recombinant antihemophilic factors (e.g., Alphanate, Recombinate, Bioclate, Helixate, Kogenate) control bleeders' disorders, and various interferons (e.g., Roferon-A, Infergen, Avonex) manage diseases ranging from leukemia to hepatitis to multiple sclerosis. Epoetin (Epogen) medicates anemia associated with chronic renal failure or chemotherapy, and tissue plasminogen activator (Retavase; commonly known as TPA) treats myocardial infarction, a form of heart disease.

Other market-approved drugs from recombinant DNA technologies include Alferon N for treating genital warts, Abelcet for invasive fungal infections, Carticel for knee cartilage damage, Ceredase for Gaucher's disease, and Pulmozyme for cystic fibrosis. The hormone erythropoietin, which stimulates the production of red blood cells and hemoglobin in bone marrow, is one of the top-selling drugs to have emerged to date from the recombinant DNA revolution, with annual sales in recent years grossing about $5 billion.

Given the efficacy of GM microbes as living factories for producing large quantities of a variety of recombinant DNA drugs, why are scientists also harnessing transgenic plants (chapter 4) and barnyard animals (chapter 5) to produce these drugs? One important reason involves the issue of gene processing.

Inserting a transgene into a foreign creature is merely a first step in successful drug production. Once inside the GM organism, that transgene must be recognized and managed by the recipient cells. This means that it must be activated properly, its messenger RNA translated efficiently, and its polypeptide product then processed (e.g., glycosylated) in ways that yield a mature protein. The last stage (known as post-translational modification) is especially tricky, and there may be as many key steps along the protein's production pathway as there are in a factory line that produces a finished automobile.

Bacteria can handle the post-translational manipulations for some kinds of transgenic proteins, but not others. Many mammalian transgenes, for example, are beyond the microbe's native processing capabilities. Genetic engineers sometimes can tinker further with a bacterium to upgrade its gene-processing skills, but it is often easier just to use a microbe or other vector to insert a mammalian transgene into GM plants or livestock (chapters 4 and 5). Such complex forms of life are more likely to come ready equipped, thanks to evolution, with the molecular and cellular wherewithal to complete the task of generating a usable protein product.

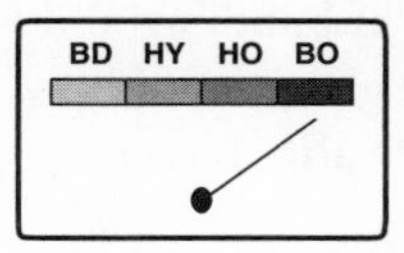

More Industrious Microbes

The pharmaceutical industry is not the first arena of human entrepreneurship to use microbes. As early as 6000 B.C., Mesopotamians in the region of what is now Iraq began using naturally occurring yeasts to clot cheese and to ferment alcoholic beverages, two activities that became quite popular through the ages. Other microbes were adopted for commercial application more recently. For example, a natural root-nodule bacterium that fertilizes agricultural soils with nitrogen was marketed commercially in 1895. In 1912, researchers harnessed microbes to produce industrial quantities of acetone and butanol in the first major use of microbial actions for market products other than food and drink. And, in 1938, in the first government-sanctioned release of a bacterial product in pest control, *Bacillus papilliae* was sprayed across the eastern United States in efforts to control the Japanese beetle.

Of course, all of the above are examples of the use of naturally occurring microbes. In recent decades, microbes also have been genetically modified for human service, often as living factories to produce enzymes of commercial importance. The seven primary areas of industrial application have been:

(a) *The food and feed industries.* Bacterial enzymes are used to process many foods including dairy products, grains, sugars, starches, alcoholic beverages, juices, meats, and so on. For example, amylase and malt from yeast are crucial in the baking industry. Microbial enzymes also are used to process certain animal feeds (such as barley meal for chickens) into more digestible and/or nutritious fodder. Typically, microbial enzymes are catalysts and convert one form of the food substance to another, but themselves are either absent or nonoperational in final products.
(b) *The cleaning industry.* Enzymatic proteases are well-known washing agents or biological detergents found in most laundry products. They help remove protein stains left by blood, sweat, or grass. Newer laundry additives include enzymes such as Lipolase that remove fat stains left by lipstick or butter and Celluzyme that softens fabrics made of cellulose fibrils. Released in 1988, Lipolase reportedly was the first industrial enzyme developed via genetic engineering.
(c) *The textiles industry.* Several enzymes purified from microbes are used in textile processing. These perform various tasks such as bundling and sealing fibers into durable threads, removing fuzz for a glossier finish, or removing harsh chemicals such as hydrogen peroxide after bleaching is completed.
(d) *The pulp and paper industry.* Industrial applications of microbial enzymes include removal of pitch and other biodegradable deposits from papermill machinery, lignin degradation during pulp production, and the treatment of starch to improve its coating performance on finished paper products. Old-style pulp operations notoriously pollute the environment, a problem that enzymatic pulp treatments can help to alleviate by reducing the need for toxic chlorine compounds that are the traditional bleaching agents used to soften wood fibers.
(e) *The leather-tanning industry.* Animal hides contain proteins and fat, sandwiched between collagen fibers, that must be removed before the tanning process can proceed. Enzymatic proteases and lipases do this.
(f) *The oils and fats industry.* Particular biological enzymes catalyze conversions between various forms of fatty acids, affording opportunities for tailor-making desired fats and oils. For example, via chemical catalysis in factories, low-grade substances such as palm oil can be upgraded to higher quality forms with a wider range of commercial applications.
(g) *Diagnostics and testing industries.* Many biological molecules exhibit high specificities that make them ideal biosensors for detecting particular organic entities, such as sugars, amino acids, peptides, cell types, or

antibiotics, in complex mixtures of otherwise unknown composition. In one such example of commercial significance, DNA kits now are available for detecting common food contaminants such as *Salmonella*, *Escherichia coli*, and other microbes that may be dangerous when ingested. Compared to earlier diagnostic methods involving microbial culturing, these new DNA diagnostic kits are easy to use, precise, reliable, and fast.

Overall, in the seven industrial areas described above, a published compilation in the year 2000 identified approximately 200 microbial biotech enzymes in the commercial marketplace, many with registered trademarks. Not all of these enzymes were produced using recombinant DNA technologies, but most (perhaps 90%) were. Here's how it's often done.

A desirable enzyme from any suitable species (bacterial or otherwise) is identified and characterized and its coding gene isolated. Using DNA-manipulating methods, this coding sequence then is transferred into *Bacillus* or *Streptomyces* (two types of bacteria), *Aspergillus* (a fungus), or *Saccharomyces* (a yeast). These microbial taxa are popular for genetic engineering applications because they are well understood genetically, safe, easy to handle, and quick to grow into vast colonies in large fermentation vats. The microbes then take up the transgene, clone it as they themselves divide and multiply, and successfully transcribe and translate the transgene into the protein product desired. Then mass cultures of the GM microbes can be grown to produce commercial quantities of highly pure protein.

The pharmaceutical factories and their fermentation vats, in addition to merely housing the microbes, can provide an additional advantage for the genetic engineering enterprise—containment. Through such practices as separating the GMOs from the desired protein before it is sold, inactivating or sterilizing the fermentation broth after use, or chemically treating any effluents from the production facility, the final commercial product can be cleansed of any contaminating GM microbes. Thus, when the facilities are operated properly, any potential environmental or health risks due to escape of the GMOs are minimized.

The last 10 years have witnessed an explosive growth in the use of microbial systems and recombinant DNA technologies to produce enzymes beneficial to industry, and often to consumers and the environment as well. Considering that fewer than 1% of the world's microbial species have been cultured and characterized and that only a tiny fraction of the latter have been further engineered by human hands, prospects for the development of many more microbial enzymes of commercial importance would seem to be tremendous.

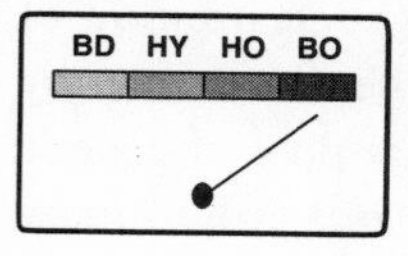

Accelerated, Directed Evolution

Despite all the fancy gadgetry for manipulating DNA in biotechnology laboratories, with few possible exceptions (e.g., see discussion of poliovirus in chapter 2) genetic engineers have yet to create, from scratch, anything so complex as a living entity or organism. How then has nature filled the planet with such inventions? It has done so through endless genetic tinkering and relentless selective filtering across billions of years of evolution. The fodder for evolution is vast genetic variety which arises continually from mutations and from recombinational shuffling of DNA (notably via sexual reproduction in most species). The directive aspect of evolution, natural selection, then sorts out this genetic variation, generation after generation, saving and cumulatively improving variants that promote survival and reproduction and ruthlessly eliminating those that do not.

The recombinational component of genetic variation is particularly impressive. Consider, for example, a typical sexually reproducing species with 40,000 functional genes (about the number that humans possess). If we conservatively assume, for the sake of argument, that each of these genes occurs in that species in only two different forms (alleles), then the theoretical total number of different diploid genotypes that could arise via recombination is $3^{40,000}$, a number so astronomical that it more than rivals the number of subatomic particles in the known universe. Clearly, this full potential can never be realized, but the point is that recombination is an incredibly powerful way to generate and display genetic variety from which natural selection may pick and choose.

In 1994, it occurred to Willem Stemmer that genetic engineers might be able to speed up the commercial development of drugs and vaccines by imitating nature's ingenuity. In an experimental test of this idea using the prokaryotic bacterium *Escherichia coli*, Stemmer digested the TEM-1 gene (which codes for an enzyme that breaks down the antibiotic cefotaxime) into small fragments and then randomly reassembled the pieces into full-length DNA sequences. He called this procedure "*in vitro* DNA shuffling," and it was indeed rather analogous to shuffling a deck of cards repeatedly. The reassorted DNA sequences then were inserted into legions of bacteria and tested for capacity to confer tolerance to cefotaxime (the selective agent). After just a few rounds (generations) of this evolution-mimicking process, Stemmer created novel TEM-1 enzymes that could detoxify a 32,000-fold higher concentration of cefotaxime than did the originals. In effect, by generating genetic variety by molecular recombination and then selectively screening that pool of genetic variation, Stemmer had orchestrated an artificial yet very real evolutionary process in the laboratory.

Several methods for facilitating such directed molecular evolution have been developed and sometimes made proprietary (e.g., MolecularBreeding by Maxygen Inc.). One directed evolution approach stems from remarkable discoveries about eukaryotic organisms in the late 1970s. Scientists found that a typical protein-specifying gene in plants and animals occurs in discontinuous pieces rather than as an unbroken stretch of coding DNA. Such coding sequences (exons) are like periodic shoals imbedded in a much longer river of noncoding DNA (introns). In the human genome, for example, exons constitute about 5% of a typical gene and introns make up the remaining 95%. It also turns out that the exons of a particular gene often specify distinctive structural domains that can be viewed as "functional cassettes" within the resulting protein.

With regard to the natural evolution of protein-coding genes, these unanticipated findings yielded important new insights: Most recombinational events probably arise in introns because they comprise the vast majority of a DNA sequence; such recombinational events shuffle the exons of particular genes; and that by reassorting the protein-coding cassettes, exon shuffling might be a hugely creative source of functional genetic novelty. Now, some genetic engineers are mimicking these natural processes by generating and screening bacterial libraries containing combinations of exons that have been artificially shuffled *in vitro*. As a result of this laboratory procedure, protein cassettes from different genes and even from different species can be mixed and matched into countless arrangements, some of which, it is hoped, will result in recombinant proteins with new or improved functions.

In terms of commercial applications, directed evolution is mostly in developmental stages, but proof-of-concept studies demonstrate dramatic kinds of genetic alterations that can be achieved. For example, artificial gene shuffling and screening in the bacterium *Streptomyces fradiae* quickly yielded a strain that produced 10-fold more tylosin (a commercially important antibiotic) than before. Likewise, shuffling of gene components among four divergent bacterial species achieved hundred-fold increases in the activities of the targeted cephalosporinase enzymes. In yet another such gene shuffling experiment, scientists evolved a novel DNA sequence for fucosidase (an enzyme that breaks down the sugar fucose) from precursor gene sequences whose products processed another sugar, galactose. The next, ongoing, step in directed-evolution research will target proteins of pharmaceutical or industrial importance, the intent being to engineer and then microbially mass-produce molecules with such features as increased potencies, novel biological activities, or longer shelf lives.

In these regards, human-directed molecular evolution offers some potential advantages over molecular evolution run by nature. First, scientists can reassort exons from the same or different species, whereas nature usually (with some important exceptions) confines genetic shuffling to recombinational exchanges among DNA sequences within a given species. Second, scientists can engineer experiments to create complex progeny containing gene parts from multiple sources, whereas nature's sexually produced progeny have only two parents each. Finally, researchers have specific design or performance goals for proteins to be biomanufactured, whereas evolution in nature faces oft-competing molecular demands in assembling and maintaining whole suites of interacting genes that promote the viability and fertility of organisms.

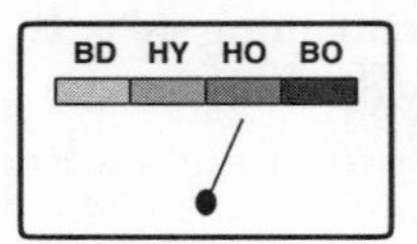

In summary, accelerated directed evolution holds considerable quiet promise that has yet to be realized. Hence, this enterprise to date falls squarely in the hope category on the boonmeter.

Concluding Thoughts

Microbes have been harnessed, and often engineered, to bioproduce many substances of value to mankind, from pharmaceuticals and therapeutic drugs to enzymes with applications in the food-processing and cleaning industries. Thus, bacteria and viruses not only enabled the recombinant DNA revolution by offering the original conceptual inspiration for genetic manipulation and gene transfer among species, but they also provide most of the molecular tools (such as enzymes that cut and paste DNA) now adopted by genetic engineers for human biotechnology purposes. Indeed, microbes often constitute the living workshops and biological factories in which scientists routinely deposit and clone various transgenes that churn out valuable proteins for medicine and industry.

All of these developments are even more remarkable given the recency of the relevant scientific knowledge. For example, bacteria and viruses were discovered little more than a century ago; DNA was not recognized as the genetic material of life until about 1950; and restriction enzymes that cut long threads of DNA into manageable-sized pieces were unknown until the late 1960s. The GM enterprise per se began only in the 1970s, yet just 30 years later its recombinant DNA products already are woven thoroughly into the fabric of human commerce in the industrialized world.

Unlike many other biotechnology applications that involve the outdoor release of GM organisms, including bacteria and viruses (see chapters 4 and 6), most recombinant microbes that are used to produce industrial or pharmaceutical proteins are housed in large, indoor vats. Hence, these GM microbes normally remain contained and pose few direct risks to human health or to ecological processes in nature. Although the transgenic proteins may leave the factory as marketable products of genetic engineering, the GM microbes themselves typically do not. For these reasons, the practice of engineering microbes for pharmaceutical drugs and industrial compounds generally has been a boon to society, with relatively few potential drawbacks.

However, another possible use of GM microbes—in biological warfare—carries consequences that are almost too horrible to contemplate. Unlike conventional physical and nuclear weapons that for better or worse remain in principle under direct human control, a disease microbe once unleashed cannot reliably be stopped as it pursues its own evolutionary agenda of self-perpetuation. Might some recombinant microbes, especially if generated or utilized by people with malice, cause great human suffering? The question must be asked, especially in our post-9/11 era of global terrorism.

One final point is that biotech pharmaceuticals and industrial products for the most part have entered the marketplace with relatively little fanfare or even public notice. This stands in sharp contrast to the situation with GM crops and farm animals (chapters 4 and 5). Although many industrial and medicinal products from GM microbes have infiltrated society and sometimes greatly benefited human lives, governments and societies continue to debate the merit of transgenic food sources produced by similar kinds of gene-manipulation techniques.

4

Getting Creative with Crops

About 10,000 years ago, our forebears invented agriculture and its sister enterprise, selective breeding. By purposefully sowing the seeds of desirable wild plants, harvesting the resulting foods or fibers, and retaining seeds from favored specimens for subsequent planting, our ancestors gradually transformed native plant varieties into the bountiful corns, rices, cottons, tomatoes, and other productive domestic crops that we enjoy today. Such artificial selection over the eons required no sophisticated gene technologies, no detailed understanding of hereditary mechanisms, no cognizance of evolutionary processes—just a keen eye for desirable plant traits, strong arms to till land and tend crops, and patience.

Today's crop scientists still need a keen eye for their subject, but great patience is no longer required. By manipulating DNA directly through recombinant technologies, modern genetic engineers are seeking to improve crop quality often dramatically and nearly overnight. As illustrated by the case studies presented in this chapter, great promises and potential pitfalls attend this endeavor as it has been applied to some of mankind's most important domesticated plants.

Before the 1960s, the notion of moving foreign genes into plant species was fantasy. Then a Belgian research team led by Lucien Ledoux claimed to have accomplished this feat by integrating bacterial DNA into barley. But the evidence was controversial, and the general approach was greeted with skepticism. In the mid-1970s, the Ledoux group asserted that by transferring a bacterial gene into a mutant strain of *Arabidopsis* (a small species of flowering mustard), they had restored the plant's capacity to produce vitamin B_1. At about that same time, a group led by Dieter Hess in Germany claimed to have engineered a change in petunia flower color by shuttling in a foreign gene. However, such possibilities for "horizontal" gene transfer (gene transmission other than from parents) in plants were not broadly acknowledged by agronomists until about 1984, when the scientific evidence for such gene-engineering events finally become incontrovertible.

The Monsanto Corporation was one of the first agribusinesses to nurture and exploit the nascent field of plant biotechnology, and it soon became a major player in the burgeoning field of crop genetic engineering. Before 1990, little land was devoted to transgenic crops, but that situation changed dramatically within the decade. The first GM crops were approved for commercial planting in the early 1990s, and by 1999 more than 40% of the corn, 45% of the soybean, and 50% of the cotton planted across the United States was genetically modified. Worldwide in that year, 70 million acres (roughly the size of Colorado or of New Zealand) were devoted to transgenic crops, and the total has continued to rise. Now, GM plants cover more than 130 million acres—roughly the area of Spain. Several countries including Brazil, Argentina, and Japan have promoted transgenic crops aggressively, and a recent report in the journal *Science* presented conclusions that transgenic cotton has significantly improved the lot of small farmers in China, as well as increased that nation's efficiency in crop production.

On North American grocery shelves, 70% or more of processed foods now contain ingredients from genetically engineered crops. To date, most of the transgenic plants under commercial or experimental cultivation have been engineered for one or another of the following: genetic resistance either to microbes, insect pests, or herbicides; improved food quality; or as live biological factories for producing pharmaceutical drugs.

Consumers in the United States generally have accepted this transition to GM crops with little notice. In the United Kingdom and continental Europe, however, public outcries against transgenic "frankenfoods" and agricultural "farmageddons" have been loud and persistent. Governmental responses in Europe have varied, but France and Britain, among others, have opposed transgenic crops rather vehemently; and in 1998, the European Union enacted a four-year moratorium on further introduction of transgenic foods. Divergent public sentiments on the two sides of the Atlantic serve notice that apart from the science per se, societal attitudes and political and economic considerations (such as agricultural protectionism) will play a huge role in the commercial acceptance or rejection of recombinant DNA crops.

The available scientific evidence often leaves room for polarized opinions on many GM crops, especially with regard to environmental and health issues. Among the potential blessings of GM plants are increased crop yields per acre, nutritional and medical benefits to domestic animals or humans, and some exciting ecological payoffs such as diminished needs for tillage (which can promote soil erosion) or for poisonous chemical pesticides. With regard to alleviating world hunger, some prognosticators think that transgenic plants will have even greater impact than did the novel genetic crop strains developed during the Green Revolution, begun in the 1950s, by aggressive use of more traditional plant-breeding methods.

However, there are important concerns. Some GM plants may pose risks to human or animal health, such as in generating allergic responses. Some transgenes might escape to nontarget plants of the same or different species and perhaps yield unwelcome ecological outcomes, such as conferring herbicide resistance to weeds (undersired plants). Environmental damages of several sorts (discussed later) could stem directly or indirectly from agricultural systems devoted to transgenic crops. Furthermore, the wide deployment of GM crops in developing nations and industrialized countries likely will promote shifts toward large agribusinesses, with profound economic, environmental, and social consequences that are not consistently beneficial. Finally, many people simply are ethically opposed to any overt genetic tampering with food crops.

As laboratory techniques for generating transgenic organisms continue to evolve, so too do the regulatory procedures of relevant governmental agencies. In the United States, the outlines of such policies appeared in a landmark 1986 publication sponsored by the White House Office of Science and Technology Policy, *Coordinated Framework for the Regulation of Biotechnology*. This publication set forth the principle that any products of biotechnology should be monitored and regulated (as are those of other technologies), but that biotechnology as a process required no special oversight because its techniques are not inherently risky. Two subsequent white papers sponsored by the U.S. National Academy of Sciences (NAS) endorsed these concepts. They also promoted the idea that in terms of inherent properties, GMOs do not differ fundamentally from nonmodified organisms. At the National Institutes of Health, strict guidelines for producing transgenic organisms were initiated in 1976, but over the years these have become much less restrictive generally as scientific experience and confidence with GMOs grew. In the United States today, three federal agencies have primary responsibility for implementing a variety of evolving laws and regulations regarding GM crops (see box).

Regulation of Biotech Crops

Three federal agencies oversee GM crops in the United States: the Department of Agriculture (USDA), the Food and Drug Administration (FDA), and the Environmental Protection Agency (EPA). The jurisdiction of the USDA includes crop plants and their pests; that of the FDA includes human food, food additives, animal feed, plus veterinary and human drugs; and the EPA monitors areas such as plant pesticides and GM microbes. The primary roles of these three agencies can be summed up by the following types of questions (often overlapping) that each asks: Is it safe to grow? (USDA); Is it safe to eat? (FDA); Is it safe for the environment? (EPA).

More specifically, the FDA typically requires, for example, evidence that food proteins from GM crops are from sources with no history of allergic or toxic effects; do not resemble known allergens or toxins; are present at low levels and are degraded quickly within the human stomach; and at high consumption levels have not adversely affected experimental animals. FDA regulators routinely evaluate data from the following types of studies regarding bioengineered proteins: toxicity responses in laboratory animals (usually mice); digestibility under simulated human gastric conditions; comparisons of amino acid sequences against those from known allergens; heat stability (does cooking break down the protein?); and physical positions within the GM plant at which transgenic expression is high. Analogous types of oversight are performed by the USDA and EPA within their respective jurisdictional domains.

Nonetheless, critics often complain that oversight is not tight enough. They suggest, for example, that the biotech-friendly USDA mostly operates an honor system in which the GM companies themselves normally design and conduct research trials in accord with general agency guidelines. However, to be fair, this is also common practice in many kinds of regulatory agencies in the United States (for reasons relating to feasibility and governmental costs). And it is also true that food crops produced by traditional breeding methods usually receive much less regulatory attention than do GM plants.

Ironically, any public perception (right or wrong) of governmental laxity regarding GM crops would do the biotech industry no favor in the long term because it might foster increased suspicion about questionable GM practices. Without sound scientific information and trusted oversight, neither industry nor government can convincingly allay societal fears about GM foods. The NAS report published in 2000 included many constructive recommendations for improving governmental oversight of GM agriculture.

Some of the difference in European versus American responses to GM crops probably traces to cultural views about governmental roles. In Europe, a precautionary principle is the prevailing norm. Its fundamental tenet, summarized in the slogan "better safe than sorry," is that government has a proactive responsibility to protect public health and the environment even in the absence of clear scientific evidence of potential damage. In the United States, in contrast, a prevailing philosophy allows corporate science (with only modest oversight from regulatory agencies) to quantify perceived benefits and risks and then decide whether the value of a commercialization project outweighs the dangers. Given the immediate potential for GM crops to profit particular constituencies, and the delayed and more diffuse nature of any ecological or societal damages that also might be entailed, it should come as little surprise that agribusiness in North America has grown with great vigor.

Behind this exuberance there is still a sobering message: no definitive scientific audit of any major GM crop, anywhere, has tallied both the assets and liabilities of the ecological, medical, economic, and social ledgers (admittedly, this would be a truly daunting task). What exists instead are numerous partial assessments, often by vested-interest groups with underlying agendas, that consider only selected categories of potential risks and benefits. Nonetheless, enough information has accumulated to indicate that the new enterprise of agricultural genetic engineering is powerful, commercially successful, and yet in some ways a mixed societal blessing.

Combating Corn Borers

Chemical insecticides in agriculture have a tarnished history, despite their promotion as magic bullets by some commercial manufacturers. In 1948, Paul Müller won a Nobel Prize for his 1939 discovery that DDT killed insects, and in the 1950s this and other synthetic toxins saw their first widespread deployment to control agricultural pests. The initial euphoria was misguided, however, as the destructive ecological effects of this broad-spectrum neurotoxin eventually became too great to ignore. Long-lasting residues from DDT become concentrated as they move up food chains, often killing songbirds and other animals as well as insects. In the 1960s, some of North America's most spectacular native species including brown pelicans, peregrine falcons, and bald eagles were driven to the brink of extinction when DDT products accumulated in their tissues and caused massive reproductive failures. Thank goodness Rachel Carson sounded her eloquent alarm in 1962, and the United States later banned this poisonous chemical, or else we might forever be facing "silent springs."

Every year, U.S. crops are showered with nearly 1 billion pounds of various pesticides, mostly for the control of insects, weeds, and fungi. Although these synthetic toxins effectively reduce populations of many crop-destructive pests, at least temporarily, most are blunt instruments of pest control. They can devastate desirable creatures, cause health concerns for humans, promote the evolution of insecticide resistance in pest species (some houseflies already had evolved DDT resistance before Paul Müller won his 1948 Nobel Prize), wreak environmental havoc, and in general divert attention from integrative, multifaceted pest-management protocols with sounder ecological footings.

For these reasons a quest began for natural biopesticides as alternatives to synthetic chemical pesticides. This effort got a huge boost with the characterization of pathogenic strains of *Bacillus thuringiensis* (*Bt*), a widely distributed soil bacterium. Known to science since 1911, this microbe houses plasmids carrying genes that produce a wide variety of natural biological toxins, each deadly to a particular insect group. For example, protein toxins fostered by the *CryI* gene kill the larvae of some butterfly and moth species (Lepidoptera), those from *CryIII* act primarily against beetles (Coleoptera), and those from *CryIV* affect flies (Diptera). These toxins activate only in the special conditions of the insect gut, so they normally remain physiologically harmless to other creatures.

Because different *Bt* toxins kill different types of insects, they permit relatively focused biocontrol of specific agricultural pests. The traditional *Bt* agronomic products are powders or suspensions, sold under such trade names as Biobit, Raven, and Skeetal. They contain mixtures of bacterial spores and toxin crystals, applied to leaves or other environmental surfaces where the harmful insects feed. These *Bt* concoctions constitute about 1% of the total world market for agrochemicals (insecticides, fungicides, and herbicides), and they have been used in diverse applications ranging from suppression of the Colorado potato beetle to attacks on the blackfly vectors of some tropical human diseases such as river blindness (onchocerciasis).

Bt powders have revolutionized pest-control practices in agriculture, but these biological agents have some drawbacks. First, *Bt* toxins applied to external surfaces are ineffective against insects that bore into plant tissues or attack roots. Second, *Bt* powders can contaminate nontarget surfaces, affecting other insect species. Third, pest populations can evolve resistance to *Bt* toxins, especially when the latter are applied one toxin at a time and in low doses. For example, mosquitoes in some tropical countries evolved resistance to *Bt* toxins within one or two years of their widespread application, as did diamondback moths (a serious crop pest) in Hawaii and Florida. Resistance to *Bt* toxins also has emerged in experimental laboratory populations of more than 25 insect pest species.

In attempts to overcome some of these shortcomings, genetic engineers have inserted various *Bt*-toxin genes into a wide range of crops such as corn, cotton, potatoes, and soybeans. More than 100 patents now exist for such GMOs. In each transgenic crop, the production and deployment of *Bt* toxins in effect are shifted from an exogenous source (a chemical factory) to an endogenous source (the living plant itself). The *Bt* toxins expressed in the tissues of transgenic crops have proved effective against numerous pest species.

All of which brings us to *Bt*-engineered corn and its role in efforts to control the European cornborer, *Ostrinia nubilalis*. This moth species probably arrived in the United States in the early 1900s, inside Hungarian or Italian cornstalks used to make brooms. It has become North America's leading corn pest, reducing total yields by as much as 25% and causing annual losses exceeding $1 billion. Uncontested, the moth larvae bore into a plant's tissues, breaking stalks and producing poor cob growth and dropped ears. However, a larva that takes even a few bites from a *Bt*-transgenic plant soon dies of intestinal obstruction.

Bt-engineered corn offers several advantages over traditional *Bt* powders. First, the transgenic toxin is perfused throughout the plant's tissues, so it attacks cornborer larvae where they live and feed. Second, the toxin is displayed continually (although this is a mixed blessing; see below), so the time and expense associated with the reapplication of powders to the crop are avoided. Third, transgenic plants display the toxin at higher and more effective doses than *Bt* powders. So, overall, *Bt* corn can pay significant dividends to farm productivity, particularly in years of heavy corn borer infestation.

In addition to the direct production boosts, the environmental promises of *Bt*-engineered corn are staggering. Currently, about 250 million pounds of chemical pesticides—more than for any other crop—are applied annually to corn fields in the United States. To the extent that transgenic strains or cultivars of *Bt*-engineered corn diminish the need for these environmental poisons, the GM plants could be a wonderful step toward ecologically sound agricultural practices. According to the EPA, some *Bt*-engineered crops have significantly reduced the amount of chemical insecticides sprayed on the American landscape. However, a recent assessment of agricultural practices in the American Midwest revealed that corn farmers used just as much chemical insecticide on *Bt* corn as on regular corn, apparently seeing the *Bt* corn merely as a safety net in the event of a cornborer outbreak.

There are some other potential concerns as well. One is a flipside of the aforementioned benefit that *Bt* toxins are expressed constitutively in transgenic corn. Unlike *Bt* powders that have a short halflife and can be sprayed to coincide with pest outbreaks, the endogenous toxins in a field of *Bt*-engineered corn are present continually, and this can have ramifications for the evolution of *Bt* resistance in corn borers. Inevitably, insect pests exposed continually to high doses of *Bt* are under strong selection pressure to evolve *Bt* resistance. The threat is not merely academic; some laboratory populations of corn borers exposed to *Bt* proteins for several generations have evolved a capacity to withstand *Bt* toxins at levels 30–60 times greater than their nonresistant ancestors.

At the time of this writing, full-blown resistance has not emerged in natural populations of corn borers, but this eventuality is feared deeply, especially by organic farmers who long have relied on the periodic application of *Bt* powders as a highly successful means of biocontrol. If pest resistance to endogenous *Bt* toxins emerges in GM crops that are planted widely, this could compromise not only the entire transgenic *Bt* approach, but also the proven use of exogenous (non-GM) *Bt* powders as effective biopesticides.

A second concern about *Bt*-engineered corn emerged in 2000, when approximately 800 retail corn products (such as taco shells) were recalled from U.S. grocery shelves. The culprit was StarLink corn, engineered by Aventis CropScience to contain a *CryIX Bt* toxin. Because StarLink contained a transgenic protein that might elicit an allergic reaction in some people, this strain had been approved only for use as animal feed. However, the *CryIX* protein somehow made its way into the human food supply. Whether this occurred via seed-lot contamination or as a hybrid-mediated movement of the *CryIX* gene into other corn strains was unclear, but either way the economic repercussions were huge (exceeding $1 billion total). Later, the EPA ruled that *Bt* corn is safe for human consumption, but the StarLink name was tarnished and, most likely, the strain will never again be planted.

A third concern is that corn-delivered *Bt* toxins might kill nontarget insects both on and off the plant. Most available varieties of transgenic corn express *Bt* toxins in pollen as well as in ears and stalks. These pollen grains disperse widely and may be consumed, intentionally or inadvertently, by other insects. Thus, beneficial insects might be impacted by *Bt* pollen from transgenic corn, as might populations of some insect-eating birds or other insectivorous species if their food supplies thereby were diminished.

In 1999, a short research note generated much media attention. It showed experimentally that larvae of the monarch butterfly (*Danaus plexippus*) may die when fed milkweed leaves dusted by pollen from *Bt* corn. Monarchs specialize on milkweed plants, which are common near corn fields throughout the central United States. However, follow-up studies on the larvae of this and other butterfly species revealed no significant mortality effects of *Bt* corn pollen under more realistic field conditions. The authors of these latter studies concluded that any impacts of *Bt* corn pollen on nontarget insect species are likely to be negligible when compared to other mortality factors.

Despite great promise (and ardent promotion by agribusiness), it remains to be seen whether *Bt*-transgenic corn will be a lasting panacea for control of corn borers and other maize pests. Any long-term success may depend in large measure on the extent to which the approach is adopted not in isolation, but rather as one ingredient of integrated pest management.

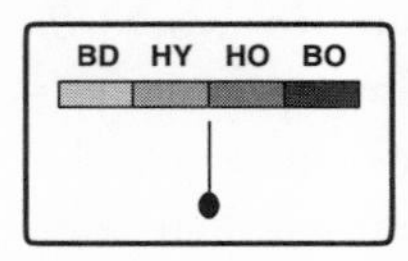

Insecticidal Cotton

Cotton is among humankind's most utilitarian and ancient of crops, as evidenced by discoveries of fossil cotton fabrics from Mexico and East India that date back 7000 and 5000 years, respectively. For more than 3000 years (beginning about 1500 B.C.) cotton was the "white gold" of India, and in conjunction with wool, it clothed much of the ancient world. In 1793, the invention of the cotton gin helped propel the commercial value of cotton ahead of wool and flax in the burgeoning textile trade of the Industrial Revolution. Today, major cotton-producing countries include India, the United States, China, Pakistan, South Africa, Australia, and Brazil.

Equally old and widespread, no doubt, are cotton-eating insect pests, including *Heliothus* budworms and *Helicoverpa* bollworm moths that can reduce crop yields by 20% and more. The quarter-inch wormlike larvae of bollworms bore into cotton bolls and feed on the plants' seeds for about two weeks before pupating. In another week or two, adult moths emerge, mate, and females lay their eggs on cotton bolls to complete their life cycle. Approximately three to five generations come and go in a growing season, with the bollworm larvae overwintering in unharvested cotton pods or in cocoons lying in soil.

Two traditional means of bollworm control have proved effective but have limitations. First are chemical insecticides, but these poisons are expensive to apply, pose ecological and environmental dangers, and often promote the evolution of insecticide resistance in target pest populations. Second, in some nearly pest-free regions such as the San Joachin Valley in California, millions of lab-reared moths, sterilized by irradiation, are released in successful efforts to stem invasions of wild bollworms. The idea is that any rare immigrant moth most likely will mate with sterile individuals and thereby be thwarted in reproducing. Unfortunately, the sterile-release method is ineffective in heavily infested areas because it requires a high ratio (60:1) of sterile to fertile moths.

As was done for corn, in recent years *Bt* genes from *Bacillus thuringiensis* have been engineered into cotton for purposes of pest management. These transgenes produce plant-endogenous toxins (such as *CryI*) that are highly effective against budworm and bollworm larvae.

The previous essay mentioned three potential downsides to *Bt*-engineered crops: human health concerns stemming from allergic or toxic effects of altered compounds in food, ecological effects associated with the death of nontarget insects, and the evolution of pest resistance to *Bt* toxins. For *Bt*-engineered cotton, the first issue is to some degree moot (cotton is not a food crop, although cottonseed oil is consumed), the second has generated little scientific attention, and the third has aroused much concern. To guard against the evolution of insect resistance to endogenous *Bt* toxins in transgenic crops, two theoretical principles of pest control have been partly implemented in modern agriculture: high dose and refuge. These tactics will be discussed in turn, with special reference to cotton and its bollworms.

"High dose" refers to the tactic of engineering plants with mixed concoctions of elevated *Bt* toxins that, presumably, should eliminate or impede the evolution of insect resistance. It supposes that any genetic changes conferring *Bt* resistance are unlikely to arise if insects are exposed to high doses of two or more distinct types of *Bt* toxins simultaneously. This assumption is often incorporated into theoretical models promoting the high-dose strategy, but recent empirical evidence has raised doubts about this concept's universal validity in the real world.

A troubling discovery was that some natural mutations in insects can simultaneously confer resistance to multiple (at least four) different *Bt* toxins, including those of the type engineered into cotton to protect against bollworms. This phenomenon was first observed in the diamondback moth, *Plutella xylostella*, but it carried a broader evolutionary lesson. (Actually, cross-resistance to multiple toxins had been known for decades, e.g., in some insects that are resistant to many chlorinated hydrocarbon pesticides as well as to pyrethroid toxins.) Subsequent research on other pest species showed one mechanism by which such *Bt* cross-resistance arises. Some *Bt*-resistance mutations act in a generic way, for example, by blocking the attachment of *Bt* proteins to the insect gut. When denied this access point to an insect's internal tissues, all *Bt* toxins are promptly and simultaneously rendered ineffective.

"Refuge," the second general pest-control tactic with transgenic plants, entails devoting some 20% of the acreage in each field to unmodified plants that nonresistant insects can attack successfully. The idea is to produce an overwhelming number of susceptible individuals each generation, so that any *Bt*-resistant insects likely would mate with the susceptibles, and thereby produce only susceptible progeny. (This also assumes that the resistance mutation is recessive such that its effects are blocked by the normal dominant form of the gene inherited from the second parent.) Thus, according to the refuge tactic, even if a particular *Bt*-resistance mutation were to arise, its spread in the pest population would be short circuited.

One practical difficulty with the refuge approach is that it requires voluntary compliance with cultivating protocols that may diminish a farmer's immediate profits because considerable acreage is sacrificed to the pest (the farmer can, however, spray the refuge with other types of insecticide). The refuge tactic also entails a hidden scientific tenet recently called into question. Researchers observed that in a *Bt*-resistant strain of the pink bollworm (*Pectinophora gossypiella*) grown on non-*Bt* plants, larvae took several days longer to develop than *Bt*-susceptible larvae. Because the *Bt*-resistant moths emerged later than normal, in the field they probably would mate among themselves rather than with *Bt*-susceptible moths, thereby violating a fundamental assumption of the refuge strategy.

Thus, from preliminary research on cotton bollworms and related pest species, two of the theoretical pillars (high dose and refuge) underlying field procedures for *Bt*-engineered cotton have been cast into partial doubt. Does this mean that transgenic cotton and other such *Bt*-engineered crops have a limited future? Not necessarily, but it does highlight the need for more basic research into the genetic bases and physiological mechanisms underlying *Bt* resistance in pest species, as well as into pest ecologies. Greater knowledge in all of these areas is necessary for understanding the likelihood and nature of any evolutionary responses to endogenous *Bt* insecticides.

Despite such reservations about the current state of development of *Bt*-engineered crops, it is important to remember downsides to the usual alternative, synthetic chemical pesticides. Potential problems fall in three main areas: public health, the environment, and the evolution of pest resistance. In humans, more than 100,000 nonfatal poisonings from synthetic pesticides are reported each year, as are an estimated 10,000 suspicious cases of cancer and assorted other public health effects. Approximately 35% of foods sampled from U.S. supermarkets have detectable pesticide residues (albeit mostly at "tolerable" levels). On the ecological front, an estimated 70 million birds die annually from pesticide exposure, together with countless billions of insects, many of which provide beneficial pollination and pest-control services. In North America alone, the economic costs of such environmental problems, though difficult to quantify, must be staggering. Finally, with regard to pesticide resistance, many insect species are well known to have evolved genetic mechanisms that circumvent lethal effects of synthetic chemical poisons (such as DDT), just as they may evolve capacities to withstand biotically engineered toxins in GM plants.

If *Bt*-engineered plants significantly reduce the world's dependence on chemical pesticides, the benefits could be huge. But as with most transgenic crops, there seem to be no guarantees. Whenever a widely cultivated plant species is modified genetically, so too are selection pressures in the field. What ensues are dynamic and often poorly predictable coevolutionary dances among the crop,

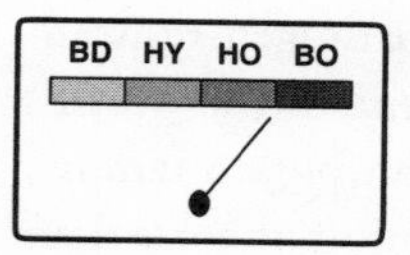

its pests, and other ecological participants. By gaining and exercising greater wisdom in both the construction and the ecology of transgenic crops and their associates, we can hope to tilt the evolutionary odds toward favorable outcomes.

Defensins and Potato Famines

Herbivorous insects are obvious pests on agricultural crops, but viruses and legions of other microscopic agents of plant diseases are at least as destructive. Thus, scientists have sought ways to engineer crop plants with genetic resistance to infectious microbes as well. In chapter 1, efforts to control the ringspot virus on Hawaiian papaya trees were described. Similar plans of attack have been drawn against other viral disease agents. For example, a transgenic strain of yellow crookneck squash, engineered for resistance to the zucchini yellow mosaic virus, has gained approval for the marketplace. Also in the research or production pipeline are many other such GM crops including transgenic potatoes and tomatoes resistant to mosaic viruses, sweet potatoes resistant to feathery mottle virus, wheat resistant to yellow dwarf virus, raspberries that can withstand bushy dwarf virus, peanuts resilient against spotted wilt virus, and transgenic plum trees resistant to plum pox virus.

Various species of bacteria and fungi cause crop diseases too, as evidenced by lesions, blotches, rots, wilting, discoloration, stunting, scabs, or tumors on plant roots, leaves, or fruits. These often reduce crop yields significantly. Nematodes (tiny worms about one-twentieth of an inch long) that feed on plant roots are another agent of infestations, responsible for global crop losses of about $80 billion annually. Many plant-associated microbes are harmless or beneficial, but the harmful ones do great damage.

This latter fact has prompted calls for biotechnological defenses, and in recent years some weapons from genetic engineering have been added to the arsenal against harmful bacteria, fungi, and nematodes. For example, genes encoding defensive proteins known as cystatins, which occur naturally in sunflowers and rice, were transferred into potatoes where in field trials they conferred on this important root crop about 70% resistance to nematode pests. Scientists have also inserted potatoes with transgenes for lysozymes, a class of proteins that promotes the breakdown of bacterial cell walls. In experimental tests, these transgenic plants displayed enhanced resistance to *Erwinia carotovora*, a bacterial pathogen that causes potato blackleg and soft rot disease. Strains of citrus, apple, and grape also have been engineered to counter a variety of bacterial disease agents.

Fungi provide compelling arguments for developing pathogen-resistant transgenic crops. First, fungi have been responsible for some of the world's most catastrophic crop failures, including the infamous Irish potato famine, beginning in 1845, that led to the starvation or forced emigration of more than 3 million people. The pathogen in that case was *Phytophthora infestans*. A century and a half later, another outbreak of potato blight attributable to this microbe affected about 160,000 acres (20%) of potato cropland in the United States. Second, traditional fungicides can be unfriendly to the environment. For example, methyl bromide (the active ingredient in many fruit and vegetable fungicides) is ecologically suspect due to its link to atmospheric ozone depletion. In the United States, the EPA, acting under the Clean Air Act, has issued regulations intended to phase out methyl bromide and other ozone-destroying substances.

One promising new approach in the war against microbial diseases is to imbue plants with transgenes producing defensive polypeptides that enhance a plant's resistance to particular fungi (or bacteria). Known as defensins, such polypeptides are produced widely in nature (e.g., in plants, insects, fish, amphibians, and mammals). In their native homes, these antibiotic peptides help protect multicellular organisms against harmful microbes. In species where particular defensins are not native, a genetic engineering goal is to insert transgenes that would arm recipients with new defensive capabilities.

In an early application of such antimicrobial genetic engineering, scientists reported in 2000 the artificial transfer of a defensin gene from alfalfa into potatoes. This transgene conferred the GM plants with greater tolerances to the fungus *Verticillium dahliae*, thereby diminishing this pathogen's virulence. Other antifungal experiments of this sort have been attempted: in banana strains that now carry a transgene for resistance to black sigatoka fungus; in transgenic rice strains that display resistance to sheath blight fungus; and in transformed barley strains engineered to carry a defensin gene from grapes that confers resistance to wheat-and-barley fungus.

What about genetic engineering efforts to repress the devastating fungus responsible for the Irish potato famine? In 1999, a research team led by William Kay (see Coghlan in references) announced the creation of GM potatoes that thwart not only *Phytophthora infestans*, but also a soil bacterium (*Erwinia*) that is a major cause of potato rot. Kay's team first spliced together a molecular construct consisting of gene sequences for cecropin (a microbe-killing polypeptide from the silkworm moth) and melittin (another microbe-killing polypeptide, found in bee venom), and then tethered these to a viral promoter sequence that switched the genes on after they were inserted into two conventional potato strains (Desiree and Russet Burbank). As hoped, the transgenic potatoes remained healthy even after exposure to

Phytophthora and *Erwinia*. Had such transgenic strains been available in 1845, the Irish crop failure might have been averted.

These antimicrobial polypeptides act by punching lethal holes in the cell walls of fungi and bacteria. Might they therefore also pose danger to humans? This seems unlikely for several reasons. First, the compounds are destroyed by cooking. Second, similar polypeptides occur naturally in many species, including *Homo sapiens*. Third, preliminary tests indicate that these substances are harmless to mice. Finally, in another context, the same two polypeptides already had undergone clinical trials as antibiotics against a dangerous superbug now lurking in some hospitals: methicillin-resistant strains of the bacterium *Staphylococcus aureus* (the causal agent of staph infections). It is ironic that consumers who might fear defensins engineered into foods might welcome them when employed as medical antibiotics.

This does raise some interesting issues, however. First, could the use of such antibiotic compounds in agricultural crops promote the evolution of microbial resistance to similar defensive polypeptides that occur naturally within the human body? No one knows for sure, although it should be remembered that defensins are abundant in nature, so microbes already must have great exposure to such compounds. Second, will microbes evolve genetic resistance to the transgenic plant toxins themselves? For GM crops, such microbial resistance might become even more common and problematic than insect resistance (discussed in the previous two essays). A third concern is ecological: Might some benign soil microbes also be harmed by transgenic defensins in crops, caught in the crossfire of antibiotic defenses aimed to shoot down pathogenic fungi and bacteria? Probably so, but again no one knows for sure.

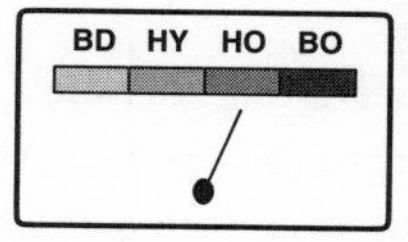

Herbicide-Tolerant Soybeans

In traditional agriculture, weed control was accomplished mechanically, by cultivation or hoeing, or culturally by crop rotation. Then, in the 1940s, chemical herbicides were introduced and "weed science" became an organized and lucrative enterprise. Synthetic herbicides are of two broad types: selective chemicals that affect only certain species or taxonomic families of plants and nonselective chemicals that kill almost every kind of plant. Herbicides were soon incorporated into standard agricultural practice, each chemical appropriately chosen to kill weeds yet leave desired crops undamaged. By dispatching weeds in cultivated fields, chemical herbicides sometimes improve crop production substantially, and, to the extent that herbicides reduce or eliminate the need for tillage as part of weed control, they can also lessen problems associated with soil erosion.

However, certain selective herbicides have several disadvantages: First, to control all weed species present, farmers often must apply two or more of these chemicals, sometimes repeatedly, during the crop-growing season, and many of the selective herbicides are quite expensive. Second, selective herbicides often are rendered ineffective when weed populations evolve resistances to them (an important topic deferred to the next essay). Third, not all crops are entirely tolerant of the selective herbicides applied to them and may suffer subtle damage that nonetheless significantly lowers crop productivity. Finally, depending on their chemical composition, mode of action, and residence times in the soil before breakdown, selective herbicides can be environmentally unfriendly and even pose risks to human health.

Nonselective herbicides also have their limitations in agriculture. Because they kill virtually all plants, these chemicals can be applied only before planting or after harvest of the desired crop, making control of many weedy species impossible. Another problem is the evolution of herbicide resistance in weed populations. However, by virtue of their broad-spectrum powers, nonselective herbicides require fewer applications than selective herbicides and can be less expensive to use. Further, the two most widely used nonselective herbicides, Roundup and Liberty, leave no long-lasting toxic residues in the soil (they are biodegradable) and appear to be safer and more benign to the environment than many other herbicides currently available.

Recognizing the potential economic and environmental advantages of such nonselective herbicides, in the mid-1990s crop geneticists began to harness yet another powerful tool in weed management: herbicide-tolerant transgenic plants. One of the earliest technical and commercial successes was the Roundup Ready soybean, engineered by Monsanto Corporation to be tolerant of Roundup, their top-selling nonselective herbicide. The active ingredient in Roundup is glyphosate, a simple compound that kills plants by disrupting their metabolism. (Glyphosate is also toxic to animals, but only at doses far higher than those sprayed on crops; thus, glyphosate use entails minimal hazard and low risk to consumers.) Roundup Ready soybeans were engineered to contain a gene, originally isolated from a *Salmonella* bacterium, that blocks the molecular action of glyphosate and thereby allows the plants to survive exposure to Roundup.

The motivation for this endeavor, of course, was that Roundup then could be applied to kill any weed species in soybean fields planted with the transgenic seed. Many farmers embraced the approach, such that by 1999, just four years after its introduction, Roundup Ready soybeans had been planted on 37 million acres in the United States (including about 44% of the Midwest's soybean acreage) and on 59 million acres worldwide (notably in Canada, Argentina, Japan, and Romania). Soon, other transgenic herbicide-tolerant crops (HTCs) including strains of cotton, corn, and canola were engineered and widely planted.

Critics of current transgenic-HTC approaches see several practical difficulties (apart from the evolution of weed resistance). One is "yield drag," a loss of productivity in transgenic as compared to nontransgenic crops. This phenomenon is common (but not universal), and its causes are often incompletely understood. Sometimes yield drag reflects the pleiotropic action of the inserted gene (i.e., unintentional effects of the transgene on plant traits other than, in this case, herbicide tolerance). For example, some strains of glyphosate-resistant soybean have a high incidence of stem splitting reportedly due to pleiotropic effects of the transgene. In other cases, yield drag results from plant-breeding practices that focus on propagating the transgene while at the same time neglecting improvement in other desirable plant traits such as vigor, nutritional quality, number of flowers, proper flowering time, and so on. More generally, widespread yield drag is a potential risk whenever an agricultural shift occurs from locally adapted crop varieties to vast plantings of a few commercially available GM strains.

The topic of yield drag is not merely esoteric. For example, a survey of Iowa farmers in 1998 revealed that although their use of glyphosate-based herbicides in conjunction with transgenic soybeans reduced weed management costs by 30% compared to conventional control measures, their net income per unit of land remained the same. This was because the dollar savings in herbicides were offset by diminished crop yields: about 4% lower in the transgenic versus nontransgenic plants. In general, if a plant invests some additional fraction of its energies toward avoiding a particular selective agent, such as herbicides, it is likely to have that much less to invest in growth or seed production or toward fending off other ecological challenges.

A second concern is that companies sometimes rush to market the seeds of transgenic plants before these herbicide-tolerant strains have been adequately tested in a variety of environments they are likely to encounter in the field. For example, transgenic soybeans raised under warm conditions tend to lose glyphosate tolerance and show an increased incidence of stem splitting.

Another issue surrounding transgenic HTCs is of monetary concern to herbicide manufacturers and agritech companies, which increasingly are one and the same. Any economic advantage enjoyed by a company, such as Monsanto, could be lost when relevant patents expire on its profitable herbicides. However, additional genetic engineering might come to its rescue. One way that a company might stay one step ahead of the game is to further engineer their herbicide-tolerant crops by inserting another gene (an inducible promoter) that activates the herbicide-resistance gene only when the plant is exposed to a specific chemical compound. To make money, the company then could add this particular chemical to a new and proprietary herbicide.

A final and much broader societal concern regarding transgenic HTCs is that they probably further promote monocultural planting practices. These include emphases on fewer major crops and plant cultivars, thus narrowing the agricultural gene pool, plus greater reliance on mass-applied pesticides and fertilizers as well as herbicides. Such practices are typically associated with conversion of heterogeneous croplands to low-diversity agrosystems. Ecological studies suggest that biotically depauperate ecosystems are less resilient to stress and more vulnerable to perturbation than complex, buffered systems. Natural biocontrol of weeds, insects, and diseases, for example, is often notoriously poor in monocultural plantings. A growing scientific sentiment is that diverse agricultural systems, with multifaceted and integrative protocols for pest management and soil-erosion control, might be economically as well as ecologically sounder over the longer term and far more sustainable.

From this ecological perspective, many of the precepts of large-scale agribusiness already were fundamentally flawed, and the wide deployment of crops engineered for herbicide tolerance may worsen this basic predicament. As stated recently by Liebman and Brummer, two scientific commentators on modern agribusiness practices, "Genetically engineered crops didn't create the mess in which American agriculture finds itself, and they aren't going to clean it up."

Because immediate benefits of herbicide-tolerant GM plants have been ballyhooed by the biotechnology industry, and because potential difficulties over the longer term mostly neglected, I have chosen to depict the overall boonmeter evaluation of transgenic HTCs in the hope category. Notwithstanding the commercial success of this GM endeavor to date, only time will tell whether the approach ultimately lives up to its huge broader promise for society.

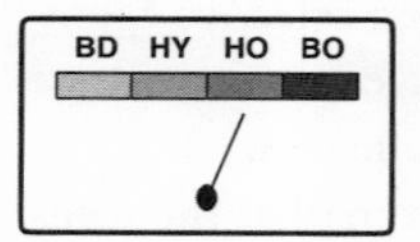

Herbicide-Resistant Weeds

No farmer grows weeds intentionally, but they all do so inadvertently. Many opportunistic weeds are the beneficiaries of a farmer's efforts to till the soil and to water and fertilize crops; hence the perceived need for weed-specific herbicides. But just as many weeds may capitalize on widespread cultivation regimens, so too, ironically, do some weed species profit from the widespread application of chemical herbicides. This occurs in two basic ways. First, weeds that already possess natural tolerance to particular classes of selective herbicide may flourish when susceptible competitors die. Second, chemical herbicides frequently promote the genetic evolution of herbicide-resistance adaptations in weed populations that formerly were susceptible.

Indeed, in recent decades, herbicide resistance has arisen in nearly 150 species of herbaceous pests from dozens of countries around the world. In one weed species or another, resistance has been reported to at least 16 different classes of herbicides, including those with such impressive names as arsenical, benzonitrile, chloroacetamide, and aryloxyphenoxyproprionate. The evolutionary response by weed populations to these chemical witches' brews is analogous to how many insect populations respond to chemical insecticides, and such evolvable tolerances to herbicides likewise pose major agricultural challenges.

Transgenic crops with genetically engineered resistance to nonselective herbicides were promoted with great fanfare beginning in the late 1990s. They provided, it seemed, the ultimate solution. With desirable plants made tolerant to herbicides such as Roundup or Liberty, two of the most lethal and broad-spectrum herbicides available, such chemicals could be sprayed at will to destroy any and all weeds that might appear on cultivated land. This idealized dream, like so many in crop genetic engineering, thus far has proved to be part visionary reality and part pipedream.

Notwithstanding the well-known capacity of weeds to evolve resistance to selective herbicides, until recently it was widely supposed that resistance evolution to glyphosate (the active ingredient in Roundup) or glufosinate (the active compound in Liberty) would be essentially impossible. After all, these chemicals are fatal to nearly all kinds of plants they touch, their modes of metabolic disruption were thought to be extremely difficult for plants to overcome, and these herbicides had been used widely for years without promoting evident resistance in wild or cultivated plant species. But, as usual, such prognostications underestimated nature's ingenuity. Reports soon appeared that populations of some weeds, such as rigid ryegrass and goosegrass, had evolved high resistance to glyphosate.

Of even greater concern, however, is that herbicide-resistance genes recently engineered into several widely planted transgenic crops might find their way into related weed species or subspecies via natural introgression (the movement of genes from one species to another via hybridization; i.e., when hybrids are formed and then mate with members of either species). For example, cultivated rice can hybridize with feral red rice, which is a serious weed pest in many of the world's rice-growing areas. Many crop species including oat, wheat, rapeseed, sunflower, millet, potato, tomato, and sorghum have close weedy relatives with whom they can hybridize. Thus, gene trafficking via introgressive hybridization is a potentially broad avenue by which particular weed species might gain unprecedented access to human-engineered herbicide resistance. This is also an avenue that basically did not exist in the pre-GM era of physical herbicide application.

To thwart the evolution of herbicide tolerance by weeds, the traditional recommendation is to apply, either sequentially or concurrently, two or more herbicides with different modes of action. This assumes that if evolution of resistance to any one chemical is a low-probability event, then the likelihood of joint evolution to multiple herbicidal chemicals is infinitesimally small. This assumption could be violated in several ways, however, if, for example, different types of herbicides share elements in their physiological modes of action that might be overcome by a common-denominator genetic alteration. In fact, many plant populations have evolved a genetic capacity to withstand multiple herbicides. One extreme example involves a weedy population of *Lolium rigidum* in Victoria, Australia, which in a scant 21 years became resistant to 9 different classes of chemical herbicides.

The growing realization that plants can evolve herbicide resistance quickly and by several routes has prompted another agri-engineering recommendation: Do not stack herbicide-resistance genes. "Gene stacking" refers to designing a given GM crop with two or more herbicide-tolerance, or other, transgenes simultaneously. For example, one variety of maize (appropriately named G-Stac), produced by the Garst Seed Company, simultaneously contains genetic modifications for resistance to Liberty herbicide, for resistance to imidazolinone herbicides, and for production of insecticidal proteins to control European corn borers. European regulations forbid this practice, the fear being that such genetic capabilities might unintendedly transfer, via hybridization, to wild relatives of the GM crops, perhaps even creating new superweeds.

Thus, considerable consternation followed a surprising announcement regarding experimental sugar-beet plots in Britain, France, and Holland. In greenhouse accidents, transgenic pollen from a glufosinate-tolerant strain of sugar beets had fertilized transgenic eggs from a glyphosate-tolerant strain, producing plants that were jointly resistant to the two most powerful and lucrative nonselective herbicides on today's world market. If this could happen in controlled greenhouses, what was to prevent its happening in nature?

Still, the dangers associated with transgenic HTCs should not be overstated nor their potential benefits minimized. For example, even if some herbicide-tolerance transgenes were to escape from cultivated crops, there is no reason to suppose that recipient weeds would be at a fitness advantage except where the herbicide in question is applied (indeed, such hybrid weeds in most ecological settings would probably be less hardy than nonhybrid wild types). Thus, the transgene would be unlikely to spread beyond agricultural fields, and the local farmer (rather than the environment and/or society at large) would bear the cost of any diminished weed control at his or her site of herbicide application. Furthermore, HTCs could provide huge ecological benefits if they lessen the need to till soils for weed control or if they reduce agriculture's reliance on more dangerous chemical herbicides. One example of the latter is atrazine, 30,000 tons of which are applied to U.S. soils each year despite strong suspicions that this herbicide might have hormonal effects that disrupt sexual development in some vertebrates.

In summary, HTCs are not likely to be agriculture's magic bullet, any more than are transgenic crops engineered for insect resistance. In both cases, what is truly altered, apart from the plants themselves, are ecological selection pressures in agricultural fields. In both cases, farmers may reasonably expect evolutionary-genetic responses by pest species, and such responses may constrain any universal or lasting benefits that these engineered crops otherwise promise.

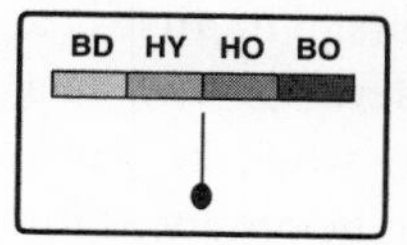

Terminator Technology

My greatest satisfaction is not in having power over people, but in having power over nature. There was a wonderful pleasure in understanding the rules of nature and, having understood them, making those rules work for me.
—Howard Schneiderman, *former leading scientist at Monsanto*

In the early years of plant biotechnology, and continuing today, major genetic engineering companies such as Monsanto had difficulty, to say the least, in convincing potential customers about the benefits of GM crops. In Monsanto's case, the reasons are many: The company often projected an arrogant attitude; it failed to appreciate, much less cater to, consumer sensibilities, particularly in Europe; for many years, its primary commercial transgenic products (plants with resistance to pests or herbicides) carried no direct tangible benefits for grocery shoppers; and it often undersold the ecological value of its *Bt*-engineered crops in reducing the need for chemical pesticides (probably because Monsanto was first a chemical company). Certainly, Monsanto's public image was not helped by the company's long involvement in producing noxious compounds. These included Agent Orange, a defoliant used in the Vietnam War, and for which military veterans later won a $180 million settlement from Monsanto for suspected health damages; and polychlorinated biphenyls (PCBs), carcinogenic chemicals that Monsanto produced from 1935 to 1977, before they were banned in the United States.

Suspicion of motives, as well as products, of agritech companies runs deep in many circles and has contributed to an active anti-GM movement around the world. In 1998, even some anti-GM activists were shocked by what they interpreted as further evidence of the industry's remarkable avarice. The occasion was the issuance of U.S. Patent No. 5723765, with the innocuous title "Control of Plant Gene Expression." Awarded jointly to D&PL (Delta & Pine Land Company of Scott, Mississippi) and the USDA, this patent gave intellectual property rights to a complicated new recombinant DNA technology that soon became known as "the Terminator." What the Terminator terminates is a plant's capacity to reproduce. What it also destroys, protestors contended, is a farmer's fundamental right of control over the age-old practice of saving some seeds for subsequent plantings.

The Terminator technology involves three components. The first is a transgene for producing a toxin that kills seeds late in their development. The second is a method to ensure that this toxin gene remains silent while the plants are under the manufacturer's control, so that viable seeds can be produced for sale. The third is a method to activate the toxin gene after the engineered seeds have been sown in farmers' fields. The manufacturer can accomplish this feat by exposing seeds it will market to a specific chemical that subsequently activates the seed-toxin gene in the farmers' maturing plants.

What prompted this endeavor to make plants sterile? Suppose that a commercial seed company develops a transgenic crop with marketable characteristics, such as high resistance to insect pests or herbicides. The addition of a patented Terminator gene to that strain then would give the company full control over the production, licensing, and sale of fresh viable seed to farmers in each planting season. Any surreptitious use or piracy of high-tech crop seeds would be eliminated.

Actually, Terminator was not the first genetic technique for making agricultural crops effectively suicidal. Most maize farmers in the United States purchase hybrid seeds annually from commercial outlets. These hybrid plants, from traditional genetic crosses between selected inbred corn lines, possess desirable combinations of traits such as high yield and resistance to particular diseases, but they also show low fertility. Any farmer who sows seeds from his first-generation hybrids would be rewarded with little or no germination in the next generation; and if some plants did grow, they would often display a common genetic phenomenon known as hybrid breakdown, in which later-generation hybrids display poor growth or survival.

In principle, what the traditional hybrid approach has done for maize production, the terminator technology can do for any crop. According to detractors, this means increased profits for seed companies, and a greater likelihood for the development of agribusiness monopolies and associated monocultural farming practices. Farmers would be rendered impotent in perpetuating the transgenic crop beyond the initial generation, and in effect held hostage to the company's engineering expertise. Agritech companies respond that protection of their marketable products (through patents and terminator technologies, for example) is necessary to cover the expense of engineering transgenic plants with desirable properties that benefit farmers and consumers. Without some such protection, they claim, the incentive for genetic improvement of crops via biotechnology would vanish.

Proponents emphasize a second major advantage of the Terminator technology: it should reduce the risk that transgenes engineered into crop plants move to unintended locales. With sterility expressly designed into the seeds of transgenic plants, some of the potential for gene movement would be blocked automatically, thus partially alleviating one of the primary general concerns about GM crops (see other essays in this chapter).

Nonetheless, critics of the Terminator technology, including the Rural Advancement Foundation International (a farm-advocacy organization) remain opposed to the Terminator. The underlying motive for this genetic technology, they say, is to enhance company profits by increasing corporate control over farming operations.

In 1998, Monsanto considered the acquisition of D&PL. This raised an even greater cry from the opponents of Terminator technology, and Monsanto backed off, even going so far as to announce its own self-imposed moratorium on the commercialization of Terminator methods. Undeterred, D&PL and the USDA forged ahead, and in 2000 the first experimental trials of the Terminator technology were conducted on tobacco plants at a USDA laboratory in Lubbock, Texas. Similar experiments involving Terminator are underway for a variety of crops, including cotton and wheat.

Despite this continuing research effort and at least some potential for ecological benefits from genetic sterility methods, the black eye that the biotech industry has received from society for promoting the terminator technology necessitates that this endeavor be labeled a boondoggle to date.

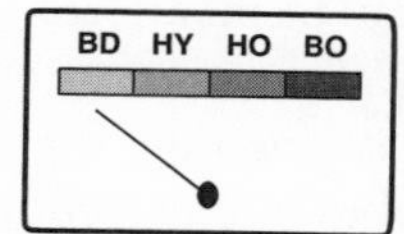

Chloroplast Concoctions

One major ecological concern about GM crops is that their transgenes might spread via gene flow into nontarget populations, or even introgress into related species via hybridization. Among the 60 major crop plants under widespread cultivation, at least 49 (82%) have wild and often weedy relatives with whom they have the potential to cross. Many nontransgenic crop plants hybridize occasionally with native relatives, and there is no reason to suppose that transgenic plants would not do likewise. Through such gene movement, the genetic uniqueness or even survival of some nondomestic populations could be compromised. In certain situations, the escape of crop transgenes into noncultivated plants could be especially undesirable. For example, superweeds might emerge if transgenes for herbicide tolerance or disease resistance were to end up in wild populations.

One such recent scare received wide media attention. In a prestigious scientific journal, *Nature*, researchers reported the detection of transgenic DNA from domesticated corn in ancestral landraces of maize in the remote mountains of Oaxaca, Mexico. Supposedly, this genetic pollution of native maize resulted from hybridization-mediated movement of transgenes from engineered U.S. corn that had been unlawfully planted in the area. This was a sobering finding because this region of Mesoamerica was the evolutionary birthplace of corn several thousand years ago, and today it remains home to the world's greatest maize genetic variety. Are the genetic purity and diversity of the native maize races in jeopardy from hybridization with domestic GM corn? Subsequent papers seriously challenged the empirical evidence for introgressive hybridization in this case, but the original authors defended their claim. Regardless of the final truth in the matter, the whole episode did serve to raise consciousness about the potential for transgenes to escape into native plant varieties.

Some biotechnologists, too, are concerned about genetic contamination of wild populations and are trying to do something about it. One approach has been to engineer ways to simply detect the movement of transgenes in nature; for example, by genetically tagging GM plants such that they or their pollen grains light up with transgenic green fluorescent proteins (GFPs; see "Reporter Genes" in the appendix). Pollen grains are the primary agents of gene flow in most plant species, and those from GM plants can be monitored more directly and reliably when visibly marked by GFPs.

A second experimental approach goes one step further by engineering crop plants with sterility or terminator genes. These are intended to block reproduction by GM plants and thereby halt the unintended flow of transgenes entirely. Several related gene-containment tactics have been envisioned but not yet applied to commercial crops. These include genetic tampering with fruiting patterns, with reproductive modes (for example, promoting vegetative propagation or asexual seed formation), or imbuing crops with transgenes that compromise the survival or fertility of hybrids.

A third general approach, involving chloroplast (cp) DNA, is especially promising. Housed in the cytoplasm of plant cells, chloroplasts are miniature organelles that play a key role in photosynthesis. Chloroplasts house small circular genomes (cpDNA molecules) that are proving to be desirable genetic engineering sites. Why might genetic engineers favor cpDNA as a location to insert transgenes? One main reason is that, unlike nuclear DNA, which is transmitted to offspring via both pollen and eggs, cpDNA in most (but not all) plant species is inherited in maternal fashion. Pollen carries few if any cpDNA molecules, so cpDNA transmission across plant generations typically occurs via female-produced eggs. This means that pollen grains, which often are carried for great distances by wind or by pollinating animals, are nonetheless ineffective avenues for the spatial dispersal of transgenes engineered into cpDNA. It also means that pollen grains from such GM plants would not carry any transgenic compounds that otherwise might potentially harm nontarget species such as butterflies (see "Combating Corn Borers").

Recently, a number of transgenes have been inserted experimentally into the cpDNA molecules of various crop species. These include particular transgenes for herbicide tolerance, for resistance to insect pests, and for plant-mediated production of various compounds with potential industrial or pharmaceutical applications. The notion is that these chloroplast-housed transgenes will be unable to escape into wild populations via pollen movement.

As a molecular site for plant genetic engineering, cpDNA has another potential advantage over nuclear DNA: it normally occurs in hundreds or even thousands of copies per cell. Thus, at least in principle, some cpDNA-carried transgenes might be expressed at very high levels in plant tissues. In one such example, researchers inserted the human gene for somatotropin into tobacco plants and found that the GM plants produced this protein in remarkably high concentrations (300-fold greater than transgenes similarly placed in the plants' nuclear DNA). Somatotropin is a valuable therapeutic compound for treating hypopituitary dwarfism in children (see chapter 3). Results from the tobacco experiments highlight the potential of cpDNA as an efficient vehicle for mass-producing pharmaceutical proteins from GM plants.

Nevertheless, several limitations and technical obstacles mean that cpDNA is unlikely to become a panacea for plant biotechnology. In at least some important crop species, such as alfalfa and possibly rice and peas, cpDNA heredity is not entirely maternal, so the absence of transgene flow via GM pollen cannot invariably be assumed. And, even when pollen is cpDNA-free, the seeds from GM plants would remain as another source of dispersal for cpDNA-housed transgenes. There is also a concern that transgenic cpDNA sequences might transfer occasionally to the plant nuclear genome, where they could be passed to progeny via traditional hereditary routes. Finally, there is some speculation that GM plants expressing transgenes at high levels in their cpDNA might grow or reproduce poorly when planted under competitive field conditions. Still, for many plant species, chloroplasts afford an intriguing alternative to nuclear DNA as a favorable target site for molecular genetic engineering.

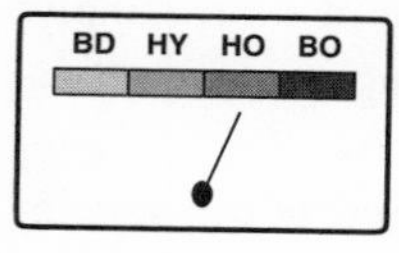

Alleviating Dietary Deficiencies: The Golden Rice Story

In addition to proteins, fats, and carbohydrates, a proper diet for the human body includes at least 17 inorganic minerals and 13 organic vitamins. Inadequate supplies of these essential substances can result in nutritional-deficiency diseases. For example, scurvy (the hallmarks of which include bleeding gums, fatigue, and eventual prostration) appears in people deprived of vitamin C for even a few weeks; pellagra, a syndrome of skin lesions and sometimes mania, follows from a paucity of dietary niacin (vitamin B_3); rickets, a disease of growing bones, ensues from inadequate supplies of vitamin D; and anemia is among the early symptoms of iron starvation.

According to the World Health Organization, one of the world's most serious nutritional health challenges is vitamin A (retinol) deficiency, a leading cause of blindness in children and an accessory factor for other afflictions including diarrhea and respiratory illnesses. More than 250,000 children in Southeast Asia reportedly go blind each year as a consequence of retinol deficiency. Worldwide, vitamin A deficiency affects the health of nearly 125 million youngsters, and it is estimated that higher doses of this essential nutrient could help to prevent about 1–2 million deaths annually.

As basal organisms in the food web, plants are the ultimate biological source of many vitamins and minerals in the human diet, supplying these nutrients to us directly, or indirectly through the animals we consume. The delivered quantity of different nutrients varies widely according to the type of plant and the mode of food preparation (storage and cooking reduce the concentration of many vitamins and minerals). For example, spinach and other leafy vegetables are a great source of vitamin K and fluoride, seed grains are excellent for biotin and several minerals, and fruits and orange/yellow vegetables (such as carrots) are rich in vitamin A. Thus, most nutritional experts recommend varied diets. But even in well-fed nations where diverse cuisines are readily available, vitamin pills and food additives are deemed important if not critical as nutritional supplements.

Sadly, in much of the developing world, people subsist on simple diets of staple foods such as corn, wheat, or rice that are poor sources of some of the essential nutrients. Approximately 30% of the world population is at constant risk of iodine and iron deficiencies, for example. Such health challenges raise a compelling new question for agrotechnology: Can staple crops be engineered to produce and deliver essential dietary nutrients otherwise in short supply? In other words, can they be genetically prefortified with higher doses of essential vitamins and minerals?

One of the earliest, most ambitious, and advanced projects of this sort involves Golden Rice, genetically engineered to deliver substantial levels of vitamin A that regular rice grains lack. Constructed in the late 1990s by Ingo Potrykus, Peter Beyer, and their colleagues, this transgenic rice cultivar, named for its golden color, soon was promoted by the media as a nutritional health breakthrough.

The science itself was a tour-de-force. Using a combination of genes isolated from diverse organisms—daffodils, pea plants, two species of bacteria, and a virus—these scientists engineered the entire biosynthetic pathway for β-carotene into experimental rice plants. This pigmented compound is a molecular precursor (or provitamin) to retinol. When ingested by people, it is naturally and easily converted to vitamin A. To almost everyone's surprise, the transgenic plants produced β-carotene, and in such quantities as to imbue the rice grains with their characteristic golden hue. The creation of Golden Rice was revolutionary in being the first case of an entire metabolic pathway being genetically engineered.

At least as complicated and interesting have been the sociopolitics surrounding Golden Rice. Potrykus and Beyer consistently have championed the view that agrotechnology must first and foremost benefit consumers, particularly the poor and disadvantaged in developing countries. But their ultimate goal of delivering Golden Rice to subsistence farmers, free of charge and without restrictions, soon encountered serious hurdles.

Most of the pathbreaking research on Golden Rice conducted in the Potrykus/Beyer lab was underwritten by public funding or by support from the Rockefeller Foundation, with neither source claiming financial stakes in any commercial products that might emerge. However, part of the research was funded under a contract with the European Commission, which included a clause conveying certain rights on project results to industrial partners of the Carotene Project. Potrykus and Beyer soon found themselves immersed in legalities and regulations—patents, technology-transfer permits to developing countries, intellectual property rights (IPRs), technical property rights (TPRs), and so on—that jeopardized their humanitarian goals. Indeed, an audit commissioned by the Rockefeller Foundation found that no less than 70 IPRs and TPRs, belonging to 32 companies and universities, had been involved in the development of Golden Rice.

In part as a concession to public opinion from the huge media attention that Golden Rice has received, some of the partnership companies recently acquiesced in various ways to humanitarian aspects of the project. For example, Syngenta, to which the inventors of Golden Rice had assigned commercial rights, announced that subsistence farmers (defined as those with annual earnings below $10,000) can cultivate Golden Rice varieties, once available, license free. In 2001, the first free samples of this experimental crop were shipped to the Philippine Islands.

Apart from the legal and economic hurdles, there are critics of the Golden Rice project who sometimes see the entire enterprise primarily as a self-serving means for agribusiness to polish its tarnished image. They point out, for example, that in the absence of a balanced diet containing fats that enable the human body to adsorb and utilize vitamin A, the vaunted health benefits of Golden Rice may elude the poorest people the crop is intended to help. They further note that a bowl of Golden Rice (100 g) probably would provide less than 10% of the U.S. recommended daily requirement for this vitamin. And they also ask whether it might not be far cheaper and easier just to distribute vitamin capsules to those in need, as indeed the United Nation's Children Fund (UNICEF) has been doing with good success at least since 1997. According to UNICEF Executive Director Carol Bellamy, a high-dose capsule, costing about 2 cents and given twice a year, is adequate to protect a child against vitamin A deficiency.

Golden Rice has been a poster-child crop of the biotechnology era, but whether it will truly help the real-life poster children of world poverty is a compelling question that only time will fully answer. Sadly, at least for now, the boonmeter must show a rating merely of fervent hope for this approach, despite its lofty goals and the power of its underlying science.

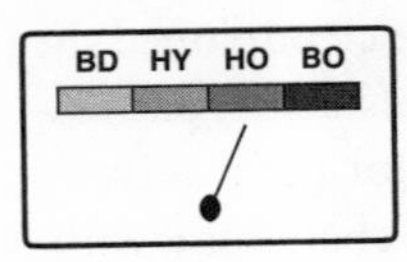

Supplementing Dietary Supplements: Nutrient Boosts

Precisely how much of each vitamin or mineral should be in the human diet has long been a matter of research and discussion. In 1941, an official body of experts in the United States published a compendium of Recommended Dietary Allowances (RDAs) that has been updated periodically ever since. For each nutrient, this book lists the daily ingestion levels judged adequate to meet the standard nutritional requirements of a normal healthy person of given age, sex, and physical condition. The adult RDA for zinc is 15 mg, for example, and that for riboflavin (vitamin B_2) is 1.7 mg. The lists were later expanded as a series of Dietary Reference Intakes (DRIs) that also include upper bounds on tolerable ingestion levels. These are important too, because there is circumstantial evidence that particular vitamins in large doses reduce the risk of cardiovascular disease and some cancers, or slow the aging process; and, on the other hand, that high-dose levels may also have some toxic effects.

One important nutrient is tocopherol (vitamin E), an antioxidant naturally synthesized only in the chloroplasts of photosynthetic organisms. The RDA for tocopherol is 7–9 mg, an amount readily obtained from a diet rich in leafy vegetables, grains, and vegetable oils. However, 10–100 times higher daily intakes of tocopherol have been associated in some scientific studies with decreased risks of various chronic disorders and degenerative diseases. Such massive quantities of tocopherol are well beyond what is consumed in a normal diet, but some scientists envision engineering food crops that will deliver these higher therapeutic levels of vitamin E.

Tocopherols come in several natural types that differ slightly in important chemical details. Two of these (α-tocopherol and γ-tocopherol) are equally absorbed by the human gut, but the α form is preferentially retained and distributed throughout the body and delivers higher effective doses of vitamin E. Unfortunately, in oilseed crops (corn, soybean, canola, cottonseed, and palm) that are primary sources of vitamin E in the human diet, γ-tocopherol is about 10 times more abundant than α-tocopherol. In the biosynthetic pathway for α-tocopherol in these and other plants, the final biochemical step involves the conversion of the γ form to the α form by the enzyme γ-tocopherol methyltransferase (γ-TMT). Thus, one notion is that if γ-TMT expression could be elevated in oilseed or other edible plants, more of their γ-tocopherol might be converted to α-tocopherol, thereby enriching human diets in useful vitamin E.

In 1998, one key research step toward this goal was achieved using an odd hodgepodge of organisms: a photosynthetic bacterium (*Synechocystis*), a microbe (*Agrobacterium*) that induces tumors in plants, a flowering mustard plant (*Arabidopsis*), and garden carrots. The following thumbnail sketch of technical events is intended merely to illustrate the molecular complexities as well as researcher ingenuity that often underlie transgenic experiments in agricultural genetic engineering.

First, the gene for γ-TMT in *Synechocystis* was isolated and used to identify and characterize the mode of action of the counterpart gene in *Arabidopsis*. This knowledge proved useful in the design strategy for a genetically modified form of *Arabidopsis* that overexpressed γ-TMT. The transgenic DNA consisted of a seed-specific promoter sequence isolated from the carrot, joined to the *Arabidopsis* gene for γ-TMT, all cloned and delivered into *Arabidopsis* by the transformation vector *Agrobacterium* (see "Galls and Goals" in the appendix). Once in *Arabidopsis*, the promoter sequence did as intended; it stimulated high expression of the adjacent γ-TMT gene, specifically in seeds. The net outcome was that much more of the γ-tocopherol was converted to α-tocopherol, and the concentration of the latter increased dramatically. As a result, the effective vitamin E activity in the transgenic *Arabidopsis* seeds was elevated by approximately ninefold.

Tocopherols and other vitamins are organic compounds, the biological products of biosynthetic pathways that in principle are potentially subject to genetic manipulations of this sort. Does this mean that inorganic minerals (composing the other major class of dietary supplements) are unsuitable for bioengineering? Not necessarily, as the following example illustrates.

Most crop plants take up minerals from the soil, but the efficiency of the process and how and where the minerals are stored in various tissues are among the variables that influence the available concentrations of these dietary supplements to humans. For example, spinach and leguminous crops are renowned for their high iron content, whereas the seeds of cereal grains are much poorer in this mineral. In an attempt to improve this latter situation, geneticists recently engineered a new strain of transgenic rice whose grains contain threefold more iron than normal rice.

The first step involved isolating the DNA sequence for an iron-storage protein (ferritin) from soybeans. Next, a regulatory DNA sequence was identified and isolated that promotes gene expression specifically in rice seeds. Then, using *Agrobacterium* as a transformation vector, this ferritin gene and its regulator were inserted into experimental rice plants. The transgenic rice has triple the iron content of its predecessor, meaning that even one meal-sized portion of the new cultivar would provide about 40% of the adult RDA for this essential mineral.

These experimental protocols for engineering the vitamin and mineral contents of plants pave intriguing paths leading to the nutritional enrichment of economically important food crops. Further research must establish whether such approaches are technically, socially, and economically feasible on commercial scales and also whether there are any possible effects harmful to human health. For example, too much vitamin D can be toxic. Also, a recent study published in the *Journal of the American Medical Association* concluded that too much vitamin A in the diet of older women may increase their risk of hip fractures significantly. One hypothesis is that too much vitamin A may inhibit the ability of vitamin D to help the body absorb calcium, a mineral needed for strong bones and for hormone production.

Another question must also be explored: Are the transgenic crops themselves negatively affected in any way by the genetic alterations? The metabolic pathways of living creatures, including those underlying the production, transport, or storage of vitamins and minerals, are marvelously tuned outcomes of millennia of evolutionary processes, so any precipitous alterations (such as the wholesale conversion of γ- to α-tocopherol, or the tripling of iron content in seeds) might have some unexpected consequences for the plants.

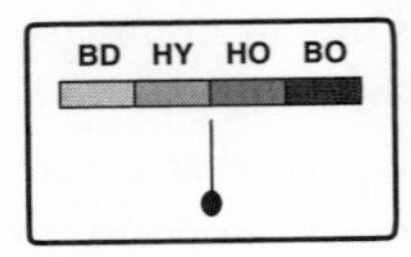

Going Bananas with Vaccines

In the late 1700s, an English physician, Edward Jenner, administered the world's first medical vaccine. Using a needle soaked in fluid from the open sore of a milkmaid with cowpox (a bovine version of smallpox that produces only mild symptoms in humans), he scratched a farmboy's arm. Upon later exposure to smallpox virus, the boy successfully resisted this otherwise deadly disease. From Jenner's experiment came not only the vaccine concept, but also the word itself, which derives from the Latin root *vacca*, meaning cow.

In general, vaccines are harmless biological agents prepared from pathogens (disease-causing organisms, typically bacteria, viruses, fungi, or other parasites) that, when delivered to a patient, elicit immune responses providing the patient with protection against the disease. The body's immunological memory varies among vaccines. Some vaccines, such as those for mumps and rubella, offer lifetime immunity, whereas others, such as for tetanus and smallpox, must be readministered after some years.

Classic vaccines are the pathogenic organisms themselves (or their close genetic relatives, as in the case for smallpox) that have been killed or seriously debilitated so as not to induce the disease during vaccine delivery. Typically, the microbes are grown in a suitable incubator (hen's eggs are often used), and then killed before being processed into a vaccine that is injected into the patient by hypodermic needle or (rarely) by oral administration. However, the entire pathogen is not required to prime the immune response. All that is needed are certain antigenic proteins, originally from the microbe's outer surface, that a vertebrate immune system recognizes as foreign. Appreciation of this fact led to second-generation vaccines that are produced from the parts (rather than wholes) of disease microbes. Such vaccines provide additional assurance that a pathogen will not somehow spring back to life when injected as a vaccine into the human body.

In the early 1990s, researcher Charles Arntzen had an epiphany of sorts: Why not engineer edible vaccines? If crop plants could be coaxed to carry and express transgenes encoding antigenic proteins of pathogenic microbes, then vaccine delivery might be as simple as eating your favorite vegetable or fruit. Infants and young children in particular would surely applaud such bio-encapsulated vaccines; far better to eat a banana or slurp down baby food than to face another needle in a doctor's office. (Actually, in the United States at least, any crops engineered to produce edible vaccines will not likely be offered to consumers directly, but rather will be orally administered by doctors; from a regulatory perspective, such plants will be considered drugs or pharmaceuticals rather than food.)

There were more serious reasons to pursue the idea. Most children in remote or impoverished regions of the world have little or no access to the conventional vaccines of modern medicine, yet this is the populace often at greatest risk for major disease outbreaks. Despite global campaigns over the past several decades to immunize all children against measles, tetanus, tuberculosis, polio, diphtheria, and pertussis (whooping cough), 20% of the world's infants are not vaccinated, resulting each year in about 2 million unnecessary deaths. Furthermore, the delivery of traditional vaccines is rather difficult and expensive even under the best of logistic circumstances. Most vaccines have a short shelf-life and must be refrigerated, for example. So, the notion of home-grown vaccines—manufactured, stored, and delivered fresh from the local garden or orchard—has wonderful appeal.

Arntzen and like-minded researchers set to work on this idea and soon reported experimental breakthroughs demonstrating the potential feasibility of this approach. For example, by 1995 Arntzen's group had inserted a protein-coding gene from the hepatitis B virus into tobacco plants, persuaded the transgenic plants to synthesize the alien protein, and showed that the plant-produced version elicited the hoped-for immune response when injected into mice. Subsequent research showed that potatoes, tomatoes, lettuce, and bananas likewise could be engineered to produce antigenic proteins from the genes of a variety of bacterial and viral species, including the causal agent of hepatitis B. Furthermore, when eaten by test animals, these transgenic tubers, greens, and fruits evoked immunological responses that conferred at least partial protection from the respective diseases when the animals were subsequently exposed to the live pathogens.

These early discoveries will be important scientific mileposts along any path eventually leading to the practical production of edible vaccines. They show that transgenic food plants can be engineered to produce and store antigenic proteins from the genes of pathogenic microbes and that when eaten these proteins often survive mammalian digestive processes to elicit appropriate immune responses. This approach is highly promising; initial calculations suggest, for example, that a single dollop of transgenic tomato paste, costing only pennies, could deliver enough antigen to serve as an effective vaccine dose. Some researchers predict that bananas will become an especially fine vaccine-delivery system for humans. Bananas grow well in many impoverished tropical regions where vaccines are badly needed; they are eaten raw (cooking reduces or destroys a vaccine's effectiveness); they do not need refrigeration; and when mashed they make great baby food. It is also possible, although much more research is needed, that transgenic plants eaten by pregnant women might result in effective vaccination of their babies.

Edible vaccines in plant foods also hold great promise for improving the health of farm animals. In 2001, results of the first such experiments were published. Corn plants were engineered to produce an antigenic protein from a common transmissible virus that attacks pigs' digestive systems and causes gastroenteritis, a major source of piglet mortality. When this transgenic corn was fed to swine in preliminary trials, the animals responded by developing a protective immunity that reduced the clinical symptoms of the disease. This study followed on the heels of research in the late 1990s in which experimental domesticated animals gained a protective immunity to the virus that causes foot-and-mouth disease when injected with antiviral vaccines produced by genetically engineered *Arabidopsis* plants.

However, many challenging questions remain about food vaccines for humans and other animals. Can plants be engineered to produce consistent levels of each desired antigen? Can improvements be made so that more of these proteins pass through the lining of the mammalian gut and thereby evoke systemic immune responses from circulating cells in bloodstreams? How often and how effectively will the ingested antigens also elicit mucosal immunity, a form of immune response mediated by tissues lining the digestive and respiratory tracts? Mucosal immunity is bypassed by traditional injected vaccines, but if it can be prompted by edible vaccines, they could offer some special advantages. For example, such vaccines could be the first against some intestinal pathogens that cause severe diarrhea and water loss in humans and kill about 3 million infants every year.

Other questions about edible vaccines remain to be answered. For example, can the vigor of transgenic plants be improved? Many of the vaccine plants engineered to date grow rather poorly when forced to produce large quantities of an exotic protein. Finally, a humanitarian concern stems from economic considerations. The primary health beneficiaries of edible plant vaccines (indeed of many transgenic crops) often will reside in developing countries, so who in the West will fund the development and distribution of GM plants to those most in need? Without lucrative market incentives, governments, international aid organizations, and philanthropists may have to step in to fulfill such goals.

In the twentieth century, traditional vaccines protected countless people against a wide range of debilitating and deadly diseases, and indeed nearly eradicated the scourges of smallpox and polio that had plagued our species since time immemorial. But not all such diseases have as yet succumbed to the approach pioneered by Edward Jenner more than 200 years ago. Perhaps soon, in the twenty-first century, a time will come when transgenic bananas, corn, and tomatoes may further improve vaccination methods, but the jury is still out.

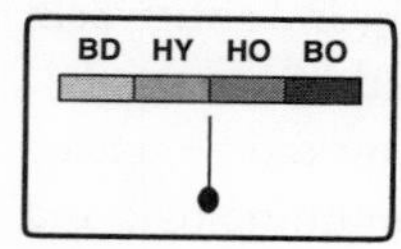

Plantibiotics and Pharmaceutical Farming

If transgenic plants can be engineered to produce microbial proteins usable as animal vaccines, could they also be genetically modified to produce human proteins of pharmaceutical or medical value? This question has not escaped the attention of biotechnologists. One of the earliest of such attempts was reported in 1997, when researchers isolated the genes specifying human hemoglobin (the oxygen-carrying protein in blood), inserted the genes into a bacterial vector, and then used the microbe to transfer the human genes into tobacco plants. About half of the genetically transformed plants manufactured human hemoglobin in assayable quantities. The basic science was impressive, and it also raised the hope that transgenic crops someday might augment traditional blood drives as a source of critical blood components. This approach would offer another advantage as well. The plant-generated blood factors, having been passed through the purifying process of transgenic insertion, would automatically be cleansed of infectious disease agents that otherwise can compromise the safety of conventional blood supplies.

For reasons of historical precedent and ease of DNA transformation, tobacco plants have been experimental workhorses for exploring many such possibilities. Since the mid-1990s, tobacco strains have been engineered to carry a wide variety of human transgenes specifying proteins such as highly specific antibodies intended to fight pathogens; therapeutic enzymes, such as β-glucocerebrosidase that could find use in relieving Gaucher's disease (a painful, sometimes deadly genetic disorder); various housekeeping proteins including hemoglobins, protein C (an anticoagulant), and somatotropin; plus an epidermal growth protein that helps repair wounds.

Other transgenic plants have been used as well. Strains of GM corn produced the first commercially important plant-derived recombinant products, avidin and β-glucuronidase (both useful in diagnostic biochemical tests). Under experimentation or development are GM turnips and rice engineered to produce human interferon for treatment of hepatitis and some cancers; transgenic potatoes to produce human lactoferrin, an antimicrobial agent; GM corn altered to produce human aprotinin, a protein that might prove useful in transplantation surgery; and GM canola plants to produce hirudin, an anticoagulant used to treat blood clots. Additional therapeutic proteins produced by GM plants include, among others, serum albumins, erythropoietin, and antitrypsin (of therapeutic potential in cystic fibrosis, liver diseases, and hemorrhages).

Depending on the type of human protein that a GM plant can be coaxed to render, two general sorts of applications are envisioned: immunologic and therapeutic. The first category of possibilities arises from the notion that transgenic plants carrying human genes for components of the immune system might yield highly specific antibody proteins (dubbed monoclonal "plantibodies") of medical utility. In particular, experimental strains of tobacco, wheat, alfalfa, and rice all have been designed to make various immunoglobulin proteins that play key roles, as antibodies, in mediating human immune responses. When harvested in suitable quantities, these proteins could find many uses in human medicine, such as in treating cancer or warding off pathogenic microbes. Potential feasibility for this approach has been demonstrated in a handful of pilot studies. For example, when a plant-secreted antibody against a bacterial agent of tooth decay (*Streptococcus mutans*) was applied to patients' teeth, it prevented recolonization by that microbe. This may have been the first reported success of plant-produced antibodies in preventing a human disease.

A second category of possibilities lies in the production of therapeutic drugs or other proteins to be administered, when needed, as additives or partial substitutes for those normally generated by the human body. The production of human hemoglobin by transgenic tobacco already has been mentioned, and in principle endless possibilities of this sort exist. Protein-based drugs of therapeutic potential in animals have been envisioned as well. For example, scientists recently inserted a dog gene for gastric lipase (a stomach protein) into tobacco, grew the plants in fields, and harvested gram quantities of the protein from the experimental crop. This gastric lipase could find application in the treatment of pancreatic insufficiency diseases in dogs and perhaps in other animals.

In theory, transgenic plants could have several significant advantages as ancillary sources for critical human and animal proteins: the plants usually have great biomass; farming technologies already exist for harvesting and processing crop plants on large scales; the processing and purification steps could be eliminated if patients obtain the therapeutic protein by eating the GM plants (as might be true for some edible vaccines); and, in some cases, production of human proteins from plants might be more economical or feasible than available alternatives, such as extracting proteins from cadavers (see chapter 3) or using large fermentation facilities for producing human proteins from transgenic microbes.

On the other hand, plant-production systems for human proteins also carry some disadvantages. First, when GM plants are grown outdoors, a special concern exists that transgenes may end up where unintended. This could happen, for example, if female gametes from nonengineered crops or their wild relatives are fertilized occasionally by transgene-carrying pollen from an engineered strain, or it could occur through physical contamination of harvested materials during storage or processing. An example of the latter occurred in Nebraska in 2002 when some cornhusks from GM maize, engineered to carry an experimental transgenic protein for the treatment of human digestive problems, were accidentally placed in a grain elevator storing soybeans. The USDA quarantined the site and ordered the disposal of soybeans valued at $2.7 million.

A second risk is that some of the transgenic proteins themselves might cause problems, either in the engineered plant itself, in animals or people that feed upon it, or in the ecological setting where the plant resides. Such possibilities should be examined critically on a case-by-case basis, and the outcome will depend largely on the nature and dose of the transgenic protein in question. Third, a major stumbling block will often be the cost-effective extraction and purification from plant tissues of therapeutic proteins that are not delivered by ingestion. Many of the transgenic plants engineered to date express foreign proteins at levels far below what is commercially competitive for routine medical or pharmaceutical applications.

Plants have been used as sources of pharmaceutical drugs across the millennia. For example, high concentrations of salicylic acid (aspirin) are found in some plant tissues that people long ago discovered could be ingested for pain relief. Some of the first "science" practiced by archaic humans probably involved the identification and study of plant species with medicinal and therapeutic properties; herbal medicine has been practiced widely (e.g., in China) for more than 2000 years. Thousands of natural medicinal compounds continue to be harvested from wild and domestic plants today or biochemically synthesized from blueprints that plants provided. However, only in recent years has modern science gained the ability to engineer GM plants that can produce therapeutic proteins from targeted human genes. It will be interesting to monitor the extent to which this new capability might revolutionize pharmacology. A tremendous promise for the future lies in this endeavor, so for now I give it a solid "hope" rating on the boonmeter.

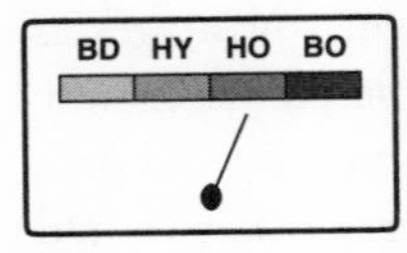

Alleviating Allergies

Twenty percent of Americans routinely sneeze, wheeze, itch, or develop skin rashes as a consequence of allergic reactions to environmental allergens—usually microbes (such as mold-producing fungi), plant pollens, or particular proteins in foods. Also common are food intolerances, which are technically distinct from food allergies. The former involve abnormal physiological reactions (often of inflammation), whereas the latter are hypersensitivities registered as responses by immunological systems. Food allergies are displayed by at least 2% of people in the United States, and food intolerances are even more common. By whatever name, negative reactions to ingested foods can be unpleasant and in rare cases life threatening.

Common allergenic foods come from animal products (notably milk, eggs, and crustacean and fish meat) and from plant products (peanuts, tree nuts, wheat, soybeans, and various fruits and citrus). Critics of genetic engineering often express concern that the protein products of transgenes, introduced into food crops, might induce allergies or intolerances in people who are sensitive to the original biological source of those transgenes. Such concerns are not merely hypothetical. In 1996, soybeans that had been genetically engineered to produce nutritious proteins from exotic Brazil nut genes became allergenic to some people. Fortunately, this problem was discovered before the transgenic soybean products hit grocery shelves.

Allergenic risks of GM foods are also taken seriously by oversight groups. For example, in 1996 a joint statement was issued by the Food and Agriculture Organization of the United Nations and the World Health Organization urging that use of transgenes from commonly allergenic plants should be avoided unless it is demonstrated that the genes do not code for allergens; that any human food found to contain transgenic allergens should not be considered for market approval unless that food is appropriately labeled as such; and that this truth in labeling should be retained through all steps of food processing and distribution. Now, as a part of formal review processes before market release, transgenic foods are tested routinely (e.g., via skin tests or immunochemical assays) to determine whether they evoke allergic responses in people or in suitable animal models.

Scientists can also apply genetic engineering methods toward reducing the allergenicity potential of commercial food crops. In the first experimental attempt of this sort, researchers in the early 1990s genetically engineered a novel strain of rice with reduced levels of amylase and trypsin inhibitors, plant proteins known to stimulate human allergies on occasion. The hope is that this may mitigate allergenic difficulties that some people face when eating rice products.

Another experimental example is provided by wheat. In regular (nontransgenic) strains of wheat, important proteins that evoke immune reactions in sensitive people are gliadins, glutenins, albumins, and globulins (listed in order of decreasing allergenicity). In the mid-1990s, researchers modified an experimental strain of wheat so that it produced extra-active thioredoxin reductase, an enzyme that influences how protein components bind together in the cell (the enzyme catalyzes the reduction of disulfide bonds). In the genetically engineered wheat, the upgraded enzyme altered the structures of the gliadins and glutenins specifically, with the net result that these two classes of proteins became significantly less allergenic (as determined in skin tests with sensitive dogs). Otherwise, the wheat plants seem to have been unaffected by this molecular tinkering.

Another current bioengineering frontier in allergen research involves the genetic modification of species producing wind-dispersed pollens. Every year, hayfever sufferers (about 20% of the population in temperate climates) face the challenge of yet another blooming season when plants release prodigious quantities of airborne allergens. Ryegrass is one notorious offender, and scientists know the primary molecular culprit: a protein called Lolp5, which is responsible for nearly two-thirds of the human immune reaction to ryegrass pollen. A few years ago, scientists genetically engineered a ryegrass strain with markedly lower levels of Lolp5 and much reduced allergenicity.

In summary, the theoretical potential for some GM food crops to express allergens has been an oft-discussed health concern, but recent transgenic research of the sort described in this essay may hold keys to alleviating rather than exacerbating the allergenicity challenges from plant proteins. However, it is still too soon to know whether this advertised goal can be broadly realized, and much of what has been published to date is little more than speculation.

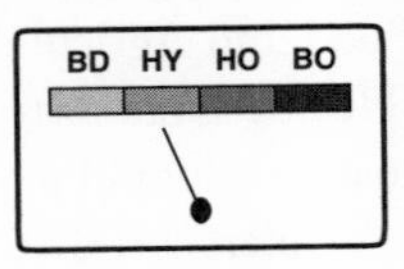

Plastics from Plants

Experimental varieties of transgenic mustard, corn, and cotton have been genetically engineered to produce the world's first plant-synthesized plastic compounds. It was hoped that such plastics, properly industrialized, might provide a green alternative to traditional petrochemical plastics such as polystyrene, polyethylene, and polypropylene, because the crop-based plastics would come from a renewable resource (living plants, rather than fossil material), and they would biodegrade relatively quickly after disposal.

The first plastic biopolymer to be coaxed from transgenic plants was polyhydroxybutyrate (PHB), a biodegradable high-molecular-weight polyester with chemical and physical properties similar to polypropylene. This polyester is produced naturally by *Alcaligenes eutrophus*, a bacterium from which genetic engineers borrowed the transgenes that were delivered to host plants. The initial experiments were conducted in 1992 on *Arabidopsis thaliana*, but the GM mustards turned out to be sickly due to overactivity of the transgenes. The first such experiments on cotton (*Gossypium hirsutum*) were conducted in 1996, with more promising results.

In native cotton, the commercially valuable fibers are seed hairs with two walls, one composed of cellulose and the other of proteins, waxes, and polysaccharides. Variation in the molecular composition and microstructure of the walls influences the cotton fibers' characteristics such as strength, thermal properties, and water-absorption profiles. In their experiments, the scientists isolated two bacterial genes (encoding the enzymes acetoacetyl-CoA reductase and polyhydroxyalkanoate synthase) that form part of the metabolic machinery by which *Alcaligenes* bacteria convert acetyl-CoA to PHB. The microbes store and use this plastic material in much the same way that animals use fats.

Using a gene gun (see "Galls and Goals" in the appendix), the researchers literally fired these bacterial genes into the cells of cotton embryos. Once incorporated and activated, the transgenes synthesized PHB from acetyl-CoA already naturally present in these plants. The resulting plastic granules, deposited between the cellular walls, measurably increased the insulating properties of the resulting cotton fibers. These were not plastic plants like artificial flowers and trees, but true living creatures otherwise showing normal growth. However, the plastic content of the seeds was minuscule—less than four-tenths of 1% of the total dry weight of the cotton fibers.

To be of genuine value to the textile industry, at least two problems will have to be overcome: First, the PHB content of cotton fibers must be increased manyfold (an outcome that might adversely affect the plants). Second, PHB naturally biodegrades when exposed to sunlight and water, so measures will have to be taken to ensure that this does not occur in any fabrics derived from the transgenic cotton. These challenges are being tackled but have not yet been overcome.

What about other potential applications? The fact that plant-produced (or microbe-produced) bioplastics are renewable and biodegradable would argue strongly in their favor, compared to their synthetic analogues made from petroleum products. Manufacturing conventional plastics consumes more than 270 million tons of oil and gas worldwide every year, depleting petroleum reserves that otherwise fuel much of the world's transportation and energy production. And, due to their abundance and resistance to degradation, traditional plastics are also a major source of environmental pollution. In 1999, President Bill Clinton issued an executive order insisting that researchers devote more energy toward finding ways to replace fossil resources with plant materials, both as raw materials and fuels.

With regard to producing biodegradable plastics from transgenic crops such as corn, the Monsanto Corporation already had been on the job for several years, but then abandoned their efforts for economic reasons. Although corn plants were engineered to produce rather high levels of plastic compounds in their stover (leaves and stems), extracting and processing these proved prohibitively expensive. Indeed, when all the financial expenses were tallied—those of harvesting and drying the stover, isolating the raw compounds with factory solvents, separating and recycling the solvents, and purifying and blending the plastics to produce a usable resin—they vastly exceeded costs of producing plastics the old-fashioned petrochemical way.

Some hidden environmental costs also made plant-produced plastics less appealing. First, with existing technologies, the processing steps listed above actually consume a greater weight of fossil fuel than is saved in usable bioplastics. Second, burning fossil fuels to produce bioplastics would contribute to greenhouse gases responsible for global warming. Thus, Monsanto and others decided it was uneconomical, for now, to pursue the production of bioplastics from GM crops. Of course, this situation could change if petrochemicals become more expensive or if governments offered economic incentives for research and development on biodegradable plastics.

A related but distinct biotechnology is being pursued by Cargill Dow, an agribusiness giant. Rather than engineering GM plants to produce plastic substances directly in their tissues, its researchers are exploring ways to manufacture new kinds of plastics from plant sugars. In 2001, the company began efforts to mass produce the first plastic of this sort (NatureWorks) that may find application in candy wrappers and other kinds of packaging.

Artificial plastic plants are manufactured and sold widely as lasting decorations for homes and cemeteries. Unfortunately, real-life transgenic plants that produce plastic compounds internally are not yet commercially successful and remain problematic as a viable economic enterprise.

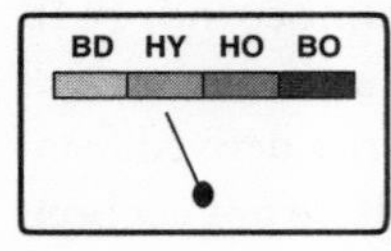

The Flavr Savr Tomato

In May 1994, consumers were introduced to genetic engineering's first ready-to-eat produce: the Flavr Savr tomato. The FDA had just ruled that the two extra pieces of recombinant DNA that Calgene Inc. had manipulated into this cultivar posed no appreciable health risks, so these GM tomatoes were as safe for human consumption as their nonengineered cousins. Actually, the FDA was not obliged by law or policy to pass premarket judgment on these plants. Rather, Calgene (later purchased by Monsanto) requested the FDA tests largely because of concerns that a genetically engineered food without FDA approval might face high consumer resistance.

What were these two extra pieces of DNA, and what marketable bonuses did they give the engineered tomatoes? One gene was merely a marker or reporter that produces a protein conferring resistance to the antibiotic kanamycin. This gene's assayable presence in a plant allowed the Calgene scientists to know precisely when their attempts to incorporate foreign DNA into a tomato strain had succeeded. Kanamycin is taken orally by some people as an antibiotic drug. After considering the matter, the FDA concluded that the small amount of the kanamycin resistance protein that Flavr Savr tomatoes added to the human diet would not affect the clinical effectiveness of the antibiotic in our species.

The other introduced piece of DNA was the real crux of the engineering endeavor. It was a gene that confers upon ripened tomatoes a resistance to spoilage. Normally, ripe tomatoes rot quickly under the action of a specific enzyme (polygalacturonase) present in the fruit. The newly introduced gene reduced the action of that enzyme and thereby delayed the fruits' bruising and spoilage. This meant that Flavr Savr tomatoes could be vine-ripened yet shipped even long distances to arrive on grocery shelves in edible condition. In contrast, regular commercial tomatoes often are harvested green to withstand shipment to distant markets, but as consumers well know, the

common result is a rather tasteless store-ripened fruit bearing little resemblance to the delicious vine-ripened specimens from their own gardens. Because many tomato connoisseurs are displeased with winter store-bought tomatoes, the hope was that the Flavr Savr would offer a scrumptious product year-round.

Unfortunately, the Flavr Savr tomato never became a market success and is no longer sold. Some consumers did not like the taste, and it was expensive. However, one benefit did ensue. The kind of basic research that created the Flavr Savr tomato, involving "antisense" technology, paradoxically has helped scientists better understood how genes are turned on and off inside plant tissues. These fundamental discoveries may be of great benefit to agricultural bioengineering over the longer term.

In engineering the Flavr Savr tomato, Calgene scientists purposefully cloned and inserted the polygalacturonase gene into the plant backwards—in reverse, or antisense, orientation. In general, such antisense transgenes inhibit the expression of their duplicate normal-polarity genes in plant cells, but the molecular mechanism underlying this phenomenon of cosuppression (also known as paramutation, or gene silencing through duplication) was a mystery. Maybe, for example, the messenger RNA (mRNA) molecules from the reverse-polarity transgene bind to the forward-polarity mRNAs to gum up the cellular works of protein translation. However, control experiments soon revealed that it did not really matter whether the transgene was cloned backward or forward; either way, the endogenous plant gene was silenced.

Further research showed that plant tissues somehow must know when too many copies of a gene are present, perhaps by sensing when the pool of RNA molecules is above some normal threshold. Under one plausible and intriguing model, viral infections provided the original selective impetus for plants to evolve a capacity to detect surplus RNAs and to silence the genes responsible. The idea is that cosuppression may be an evolved weapon originally used by plants to sense and then block the proliferation of RNA viruses that invade their tissues. Maybe the plant merely interprets the invasive duplicate transgene as just another attacking virus and silences it accordingly.

This hypothesis will require further evaluation, but regardless of the outcome, the history of the Flavr Savr tomato illustrates two broader points about GM plant research and its relationship to agribusiness. First, a successful commercial product is not the only yardstick by which the value of transgenic experimentation should be judged. Here, unexpected scientific understanding about gene regulation emerged from basic research into plant transgenesis. Second, such fundamental knowledge may prove useful in subsequent practical endeavors. In this case, future genetic engineering efforts will benefit from the discovery that simply adding transgenes to a plant's genome is not equivalent to adding functional capabilities to a plant.

Wherever and whenever cosuppression applies, by definition the insertion of a transgene for a particular protein (such as polygalacturonase) represses rather than augments the plant's endogenous production of that protein. Thus, when repression of an existing function is the desired goal (as in the case of the spoilage gene in tomatoes), the cosuppression phenomenon can be used to good advantage by genetic engineers. However, when the intent is to augment the action of an endogenous gene, the effect of adding a related transgene to plant cells has often proved to be counterintuitively counterproductive.

Despite the novel scientific insights that came from research on Flavr Savr tomatoes, the commercial product was a fizzle if not an embarrassment to the GM food industry. Thus, the project must be labeled a boondoggle with respect to its intended goal.

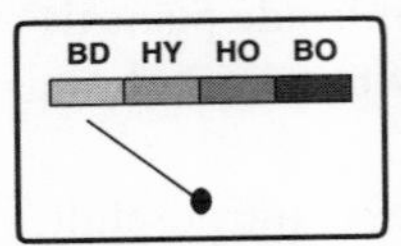

A Cornucopia of GM Products

Preceding essays have illustrated major classes of application achieved or promised for transgenic crops, but these stories are merely the tip of the iceberg of current research efforts. By 2002, at least 40 bioengineered plant strains (not invariably transgenics) were already in the marketplace, and many more will be added to retail shelves within the next few years. By recent accounts, ongoing field trials for transgenic plants number about 8000 in the United States alone and 20,000 in nearly 50 countries worldwide, involving more than 120 species in 35 taxonomic families. If even a small fraction of these trials leads to commercial fruition, consumers soon will be offered a plethora of GM foods, fibers, pharmaceuticals, and industrial products. The following are merely a few additional examples of what crop genetic engineers are up to in the arenas of cuisine, medicine, and industry.

Potatoes have been engineered with a transgene causing them to be 60% more starchy than regular potatoes, so they absorb less fat during frying. Citrus fruits are being engineered for reduced liminoid compounds responsible for bitterness. GM soybeans have been created whose oil content is shifted toward nutritious oleic acids and away from less desirable trans-fatty acids and saturated fats. Peanuts similarly modified for higher oleic acid content also appear to have a longer shelflife. Under investigation are transgenic tomatoes that may express higher levels of antioxidants, and other nutritional supplements such as lutein, which helps fight eye diseases. Experiments on GM soybeans have been aimed at alleviating an unpleasant consumer byproduct of bean consumption, flatulence.

Calgene's Flavr Savr tomato (previous essay) was a commercial failure, but this company and others have been undaunted in their quest for a more perfect fruit, and several genetically engineered strains have or soon will hit the shelves. It is hoped that Clagene's High-Sweetness tomato will live up to its name. Another example is the Endless Summer Tomato, claimed by its producer (DNAP Holding Corporation) to have superior color, taste, and texture, as well as extended shelf life (up to 40 days postharvest). Other GM produce expected to enter the marketplace in the near future may be juicier cherry tomatoes, firmer and sweeter peppers, strawberries that resist spoilage, bananas and pineapples that stay fresh longer, and new seedless varieties of eggplants and tomatoes.

The list goes on. Bean plants have been genetically engineered for improved canning characteristics such as firm texture and nonsplitting seed coats. Rice strains have been engineered to express particular proteins that prolong the grain-filling period of the plant, thereby improving crop yield. Experimental tobacco plants have been engineered to express an antigen associated with the hepatitis B virus, the purpose being to develop antibodies of use in human medicine. A commercially available strain of transgenic corn produces avidin, a protein (otherwise found in birds) that is useful to the biochemical industry in purification procedures for important molecules such as biotin.

Genetic strains of cotton have been engineered for better fiber performance. Rapeseed plants have been genetically altered to provide improved raw materials for soaps and detergents. Flowering mustard plants are under investigation that produce cellulase, an enzymatic protein used in the production of alcohol. Interestingly, the enzyme remains inactive in the living plant (and thus is unlikely to harm it), but activates when artificially exposed to high temperatures after the plant is harvested. Tobacco plants have been engineered for lowered nitrosamine and nicotine content, the goal being to manufacture cigarettes that are less harmful and less addictive.

Clearly, a huge potpourri of commodities and services is available or envisioned in the near future for foods and proteins derived from GM crops. It has been approximately 30 years since the first transgenic plants were experimentally created, and only about 10 years since the first transgenic products from agriculture were commercialized. The coming years will see a further explosion of opportunities and challenges as the field of agricultural genetic engineering grows and matures.

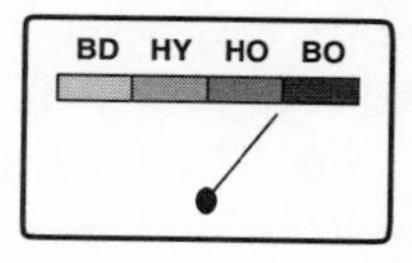

Concluding Thoughts

In a meta-analysis of the current effects of plant biotechnology on U.S. agriculture, the National Center for Food and Agricultural Policy (NCFAP, a nonprofit research organization in Washington, DC) painted a glowing picture (see Gianessi et al.). In reviewing case studies involving nearly 40 different GM crop species, 8 of which (soybean, squash, canola, and papaya, plus 2 forms each of corn and cotton) are widely planted, NCFAP concluded that transgenic plants had increased this country's annual harvest by more than 4 billion pounds, lowered growers' production costs by more than $1.2 billion, and reduced pesticide use by nearly 50 million pounds. The report also projected that if another 32 GM plants currently available or under development were likewise sowed widely, the result would be 14 billion pounds more harvest per year, $1.6 billion saved in farmers' costs, and a reduction in chemical pesticides of 163 million pounds.

The Executive Summary of the NCFAP report did not mention any potential ecological, social, or economic downsides to the GM enterprise, but perhaps this is not surprising given that the study was funded by staunch agritech supporters (e.g., the Biotechnology Industry Organization and the Monsanto Corporation, among others). However, many other reports (some presumably less biased) have also given GM crops a high rating. For example, in a 2003 *Science* article reviewing results of widespread field trials of transgenic *Bt* cotton in India, two university-affiliated authors (Qaim and Zilberman) concluded that this GM technology "substantially reduces pest damage and increases yields" (p. 900). If such experiences can be generalized, results suggest that GM crops can be of enormous benefit in developing countries as well as in the industrialized world.

Given the substantial list of transgenic crops already marketed or in the research pipeline and the impressive nature of the underlying genetic science per se (not to mention all the hoopla), it is easy to be seduced by the achievements and prospects of the GM enterprise. However, not everyone is convinced that GM crops will be a lasting panacea for agricultural problems. It is worthwhile also to reconsider thoughtful objections that have been raised by some critics, who view biotech crops more as beguiling diversions than as genuine solutions to the world's many agriculture-related challenges.

The current "Gene Revolution" in crop development may have several parallels with the earlier Green Revolution that also promised (and delivered) tremendous agricultural benefits, but not without societal costs. The Green Revolution from the 1950s to the 1980s saw agricultural productivity soar as high-yield crop varieties from improved breeding methods were introduced and farming methods were intensified. In the developing world, total wheat and rice production approximately doubled during one 20-year period, and so did the overall harvest of about a dozen important crops in the United States. Thanks in no small part to the Green Revolution, the global rise in gross agricultural output has thus far more than kept pace with the burgeoning number of human mouths to feed. For example, there is enough grain production on earth to provide, in theory, every living human with more than 3500 calories per day (nearly twice the dietary recommendation). Thus, massive hunger and starvation in the world have had far more to do with political and socioeconomic factors and distributional problems than with gross agricultural shortfalls.

In the mid-1980s, however, crop yields generally began to level off, and hidden costs of the Green Revolution also became more evident. To achieve high yields, this first generation of improved crops often required high input of fertilizers and pesticides, intensive irrigation, and heavy use of farm machinery. Unfortunately, agrochemicals polluted waterways and degraded environments, intensive cultivation sometimes depleted water resources and accelerated soil erosion, and many pest populations evolved resistance to the synthetic poisons used against them. Also, total genetic diversity in important crops was reduced as locally adapted strains were abandoned in favor of standard mass-distributed cultivars. In general, global agriculture became a more monolithic enterprise increasingly unsympathetic to traditional small-farm practices, to indigenous variety in locally adapted crops, to ecological diversity, and (some critics would argue) to long-term sustainability.

Those critics are concerned that similar directions may be perpetuated and perhaps amplified by the ongoing Gene Revolution. Although immediate per-acreage yields from GM crops often are greater, some observers think that these transgenic plants are short-term fixes at best. For example, GM plants displaying genetic resistance to insect pests or microbial pathogens will probably accelerate the evolution of countermeasures by these targeted species. This could mean a reinstated need for harmful chemical pesticides whose now-diminished use is currently one of the greatest boons of GM agriculture. Like natural (albeit somewhat simplified) ecological communities, agricultural systems are an evolutionary chessboard of genetic moves and countermoves between plants and their microbial and animal associates. In this view, GM crops are just one more way that agronomists can affect the outcome, but only temporarily.

Few agronomic experts would dispute the desirability of sustainable farming practices that meet the agricultural needs of the present without compromising the future. Nor would they dispute the typical hallmarks of sustainable agriculture: preservation of high genetic diversity in crop species, avoidance of methods that exacerbate pest problems, maintenance of soil fertility, and minimization of environmental pollution. Traditional methods for achieving these objectives include crop rotation, the promotion of diverse cultivars well adapted to local climates and soils, biological control of pests, conservation of water and soil, the use of natural fertilizers, and other low-tech farming practices. For example, recent experiments in the Georgia Piedmont have demonstrated that cover crops (such as rye or clover), when alternated with economic crops in a farmer's fields, not only suppress the growth of weeds, but also form a layer of mulch that helps protect and regenerate soils that were extensively degraded during prior decades of plowing and tilling. Thus, wiser planting practices sometimes can reap major benefits without the need for intensive herbicides and herbicide-resistant transgenic crops. The broader concern is that unbridled promotion of GM technologies might divert attention from much needed research and scientific exploration into alternative, low-tech solutions.

The Gene Revolution to date has mostly neglected or sidestepped many issues regarding long-term sustainability (ecological, social, or economic) of GM agriculture, and, indeed, some transgenic products now on the market could be interpreted as running counter to long-term goals. For example, reaping the benefits of herbicide-tolerant GM plants demands aggressive applications of chemical herbicides that typically have been produced and sold by the same companies that engineered the GM seeds. Farmers can become trapped in a revolving door of reliance on their transgenic plants, associated chemicals, and fear that weeds may evolve herbicide resistance (e.g., by the spread of transgenes from crops to weedy species). Immediate corporate profits may be high, but, detractors ask, at what price to society? Does the GM enterprise profit agribusiness more than farmers, consumers, or the environment? Roundup Ready crops and Terminator technologies are clear reminders that monetary proceeds are a prime motive underlying corporate research and marketing of transgenic crops. A realistic concern over the longer term is that the Gene Revolution may exacerbate some of the genuine social and environmental challenges associated with global food production.

Supporters of biotech crops counter that the industry can adjust its genetic engineering strategies as needs arise and as more information is gained about the risks and rewards in particular instances. They claim that transgenic foods, fibers, and pharmaceuticals are perfectly safe to use and in general are kinder to the environment than traditional crops and farming methods. They may see anti-GM activists as misinformed protestors who fail to distinguish carefully between a hypothetical hazard posed by a GM crop and the possibility of that hazard occurring (which GM supporters claim is usually remote).

Nonetheless, societies should always ask the following question about each newly proposed transgenic crop: Do the immediate and often evident improvements in the yield or quality of the GM plants more than offset the often diffuse or camouflaged expenses that may have to be borne by society at large, perhaps later? Included in these societal costs may be problems attendant with lower total crop diversity and resilience, ecological damage from the GM crops themselves or from their hybridization with related wild species (these problems are posed by non-GM crops also), a need to return to harmful chemicals if pests evolve resistance to the GM crops, increased dependency of farmers on biotechnology industries, and the social and economic consequences of possible consumer resistance (whether justified or not).

Nonetheless, there is nothing inherent in transgenic technologies that necessitates a further shift toward less sustainable agricultural methods. For example, multiple crop strains could in principle be engineered specifically to match local environmental conditions, the goal being to retain local adaptations and preserve overall genetic diversity in each crop species. Also, it is rather easy to envision how various transgenic crops now in the research pipeline might stimulate environment-friendly farming methods with lasting societal benefits. For example, nitrogen-fixing plants could diminish the requirement for nitrogenous fertilizers that pollute waterways; drought-tolerant plants requiring less irrigation could lower the demand for water (see chapter 6); crop varieties engineered for hardiness might thrive on previously degraded farmlands or in less-tilled soils; and GM crop plants with greater competitive abilities might fare better against invasive weeds and thereby lessen the need for harmful or expensive chemical herbicides. Furthermore, such GM crops could benefit small-scale as well as large-scale farms, including those in desperate regions around the world where starvation is rampant due to production shortfalls. Another arena where GM crops seem to offer great potential for societal good is in the production of pharmaceutical drugs or other more exotic compounds, like petroleum-free plastics.

In the United States, there is a profound irony underlying the current state of GM agriculture. Even as agribusinesses tout the marvels of increased crop production through genetic modification, the farm belt has been mired for decades in recession, due in no small part to crop overproduction. Farm income declined between 1960 and 2000, and the food price index for major commodities dropped by almost 50%. As farmers produced more, food supplies expanded, prices dropped, and financial losses accrued to the broader enterprise. Many small family farms, in particular, were driven out of business. To counter this vicious cycle, for decades the government has implemented crop price supports and also paid billions of taxpayer dollars in annual subsidies to encourage farming operations (usually the largest) to leave land fallow. Thus, to the extent that GM crops increase total crop or food production, they might in some ways exacerbate the economic malaise of U.S. agriculture. Still, individual farmers have no economic choice but to extract as much as they can from their land.

Where do consumers stand in the ongoing agricultural Gene Revolution? Amazingly, they have mostly been neglected to date. Many agribusinesses (Monsanto is a leading example) have focused primarily on engineering crop varieties with immediate purported benefits to farmers and, of course, to themselves. To date, relatively few transgenic products have hit the market that taste appreciably better, cost less, or are otherwise more appealing to consumers.

In general, this corporate disregard for consumer sensibilities is surprising given the economic power of the retail marketplace. Furthermore, some consumers are frustrated that the U.S. government does not mandate labeling of GM foods.

So, nearly 10 years after the introduction of the first commercial transgenic crops, it remains debatable to what extent the greater social promises of agricultural genetic engineering have been or will be realized. Certainly, it would be an egregious blunder if overblown faith in the new GM crop technologies led societies to neglect alternative approaches to relieve global problems of hunger, malnutrition, and environmental degradation that stem ultimately from human overpopulation. Perhaps it is time (finally) to devote far more effort and resources to social action on other meaningful fronts as well—family planning, public education, empowerment of women over reproductive decisions, and, in general, an exercise of greater societal wisdom on all issues related to human numbers and the environment.

Many scientific, socioeconomic, and environmental lessons should have been learned from this initial decade of experience with transgenic crops. If so, plant genetic engineering might evolve before long into a genuinely beneficial response to the broader agricultural needs of the planet.

5

Genetic Engineering in the Barnyard

In 1980, researchers microinjected pieces of DNA from microbes into the nucleus of a mouse egg, thereby pioneering the genetic transformation of mammalian germ cells by human hands. Two years later, in a similar experiment, using growth hormone genes, scientists first demonstrated that a mammalian phenotype (in this case, the size of a mouse) could be altered by purposeful genetic manipulation. In 1985, the feasibility of engineering farm animals (sheep, pigs, and rabbits) was demonstrated, and in 1987 the first transgenic poultry were produced. The ensuing years have witnessed proliferation of laboratory techniques for modifying a wide variety of genetic features in animals and also for manipulating germlines such that transgenes are perpetuated in ensuing generations. There has also been an extension from experimental research on laboratory animals to applied work on transgenic livestock, the goal being to modify the genes of barnyard animals in commercially important ways.

Another recent genetic engineering approach involves whole-creature (i.e., whole-genome) cloning. In this chapter, the word "clone" refers to any set of genetically identical organisms (or, when used as a verb, as any means by which such organisms arise). Nowadays, artificial clones are engineered directly under human auspices. They are to be distinguished from natural clones, because, long before the era of human genetic engineering, many wild creatures already exhibited a diversity of cloning mechanisms.

For example, many plants can propagate themselves clonally via root buds, stolons, rhizomes, bulbs, stem suckers, or other vegetative routes that faithfully replicate the genotype of parental individuals. Comparable means of clonal proliferation likewise are common in many invertebrate animals such as corals and sponges. Several unisexual species of reptiles, amphibians, and fishes make a routine habit of whole-animal cloning. In these parthenogenetic taxa, typically there are no males, and females give virgin birth by laying unfertilized (but often diploid) eggs that develop directly into new individuals, genetically identical to their sole parent.

Another form of natural clonal reproduction, polyembryony, occurs even in mammals. This process begins when a fertilized egg divides once or a few times in the mother's womb before embryonic development is initiated. Unlike parthenogenesis, polyembryonic clones are strictly intra- rather than intergenerational; polyembryonic siblings are genetically identical to one another, but not to their parents. Polyembryony occurs sporadically in mammals, humans included, whenever a mother conceives identical (monozygotic) twins. In one taxonomic group of wild mammals, armadillos in the genus *Dasypus*, clones arise via polyembryony in every litter.

In this chapter, however, the whole-animal clones to be discussed have been husbanded by genetic engineers rather than by Mother Nature, and their native habitat is the barnyard, not the wild. The usual intent of whole-animal cloning is, quite literally, to make genetic carbon copies of elite specimens, which sometimes are special precisely because they have been engineered to carry functionally important transgenes. Whether the salutary transgenes engineered into farm animals are perpetuated by whole-animal cloning or by conventional modes of animal reproduction, domestic GM animals promise to revolutionize farming and ranching industries.

Spider's Silk from Goat's Milk

What natural substance is stronger than iron, more elastic than a bungee cord, and able to stop speeding bullets? If biotechnologists at the Nexia Corporation of Quebec have their way, it will be a transgenic biological fiber (BioSteel) engineered from the silk genes of a spider.

With a tensile strength greater than steel, yet a lightness exceeding most synthetic fibers, ounce per ounce spider silk is among the toughest, yet pliant substances known. This amazing combination of strength and flexibility has fascinated scientists and entrepreneurs for more than a century, and numerous schemes have been concocted to harness spider's silk for applications ranging from clothing manufacture to fabrication of human artificial ligaments. The immediate technical problem has always been one of yield. How can adequate quantities of silk be harvested from small, antisocial spiders who would rather eat one another than submit to high-density commercial rearing?

Natural spider silks come in several different styles that the animals use for building webs, tethering egg sacs, swinging on safety draglines, and weaving cocoons. Each fiber type has unique properties and is composed of repetitive polypeptide units encoded by known arachnid genes. When a spider produces a thread, cells lining its silk gland secrete a proteinaceous solution that dries and is pulled into a taut crystalline cable. Scientists have attempted to synthesize analogues of spider-silk fibers biochemically but have failed. Nature's evolved process of spider-silk production has proved to be beyond what organic chemists can replicate in the laboratory thus far.

Accordingly, genetic engineers have tackled the problem by artificially splicing spider-silk-protein genes into bacteria and yeast, the intent being to coax large colonies of these recombinant microbes to produce commercial quantities of spider silk. The genetic transfers went well, but final results were disheartening: The transgenic silk proteins remained clumped together inside the microbial cells, rather than self-assembling into the desired linear silk fibers. Nor, despite repeated attempts, could the scientists artificially produce such fibers from the silk proteins that they extracted and purified from these transgenic microbes. Something about the cellular environment of these microbial organisms was inhibiting the completion of useful fiber production from the transgenic spider-silk proteins. Similar difficulties were encountered in attempts to extract usable fibers from tobacco and potato plants engineered to carry spider-silk genes.

A third avenue in the commercial production of spider silk seemed even less likely than the two approaches described above, but nonetheless has proved to be the most promising approach to date. In January 2002, scientists reported the successful transfer and expression of spider-silk genes into mammals. Using conventional recombinant DNA methods, they isolated silk genes from two species of orb-weaving spiders (*Nephila clavipes* and *Araneus diadematus*) and inserted them into the mammary cells of a cow and the kidney cells of a hamster. These experimental target cells were chosen explicitly because of their general anatomical and functional similarity to the epithelial cells in spiders' silk glands. Sure enough, the mammalian GM cells (grown in culture) incorporated the spider-silk transgenes, expressed them properly, and secreted the soluble silk proteins outside the cells where the scientists could collect them readily.

The researchers then spun the proteins into silky strands through a simple series of chemical and physical manipulations. The resulting fibers were impressive indeed—stronger than Kevlar (the synthetic material used to make bulletproof vests), but lighter and nearly as elastic as nylon. The transgenic silk was not quite as good as native spider's silk, but it represented a huge step in the right direction.

Another step by Nexia researchers was to engineer living transgenic goats carrying these spider-silk genes. They used a manipulative approach similar to that which created Dolly the sheep (see "Hello Dolly" below), but in this case the transplanted nuclei came from goat cells grown in culture that carried the spider-silk transgenes. Webster and Pete were the first two goat kids produced in this manner, and they soon were moved to a stud farm in New York so that they might sire a growing spider-silk herd. The realistic hope is that milk from the mammary glands of transgenic goats (and cows) will contain considerable quantities of usable spider-silk proteins.

Potential commercial applications abound for flexible, ultrastrong fibers. These range from syntheses of artificial tendons and ligaments to biodegradable sutures for delicate operations such as eye surgery to new lightweight body armors and high-strength composites. Another application, fishing lines and nets, could provide ecological benefits. Lines made from spider-silk proteins would decompose in aqueous environments after about a year, thereby providing a "green" alternative to the traditional synthetic nets that often pollute waterways, posing a danger to aquatic organisms as they accumulate in tangled messes.

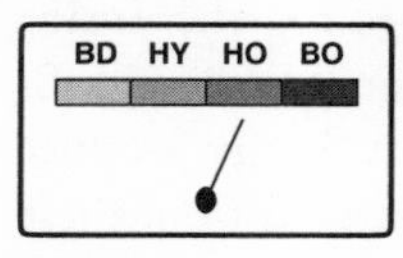

Low-Phosphorus Enviropigs

Hog ranches have gone industrial in most of the developed countries of the world and no longer bear much resemblance to storybook images of Old-MacDonald-style family farms. In a modern swine factory, huge rectangular barns house hundreds or thousands of pigs who seldom see the light of day. The animals are fed and bred indoors, and their urine and feces are washed through floor slats to be piped into waste lagoons. As these impoundments fill, surplus offal is sprayed on surrounding fields to further decompose and dissipate. Intensive hog farming increases animal production per unit of land and lowers the farmers' costs, but this often comes at great expense to the well-being of nearby property owners and to the environment.

Hog-farm odors can be overpowering, but they are only a part of the broader stink for society. Hog excrement is loaded with nutrients that can become pollutants when they find their way into streams and aquifers. This sometimes occurs in dramatic fashion. In June 1995, a large hog-waste lagoon ruptured in North Carolina, sending 25 million gallons of putrefying mass into the New River. Three weeks later, another spill of 1 million gallons spread into tidal creeks adjoining North Carolina's Cape Fear estuary. Apart from assaulting the senses, such catastrophes cause toxic algal blooms that result in massive fish kills.

Other leaks from hog farms are small or chronic, but cumulatively no less harmful to the environment over the longer term. When leached into streams or lakes, hog-farm nutrients promote a eutrophication process typically characterized by excessive growth of phytoplankton (aquatic algae) that depletes oxygen, disrupts food webs, and seriously degrades the ecological health and biotic richness of natural waterways.

One of the worst such pollutants is phosphorus (P). In the past, pigs obtained this essential dietary element from phosphate (PO_4) compounds in plants and other foods they ate. On commercial farms today, however, a primary hog food—corn seed—is a poor source of phosphorus for the animals. In the kernels of corn (unlike in germinated plants), about 70% of the phosphate is locked up as phytate, a substance that remains mostly unusable by nonruminant animals with simple stomachs, such as hogs and chickens. Thus, most of the phosphorus in phytate passes through the animals' guts to pollute the environment.

To alleviate this dietary deficiency, hog ranchers routinely supplement their animals' food with phosphate. However, this additive adds to total feed costs and likely will become more expensive in the future as humans exhaust the finite supplies of accessible phosphorus that can be mined from the earth's crust.

Thus, other strategies to deliver usable phosphorus to hogs have been attempted. One approach is to include phosphate-rich animal by-products (such as meat meal, bone meal, or processed food wastes) in pig feed, but this raises concerns about the spread of animal diseases. Another approach is to feed the swine low-phytate strains of corn (genetically modified or otherwise), thereby making more of the phosphorus bioavailable to the animals. A third practice has been to supplement hog feed with phytase, a commercially available enzyme isolated from bacteria that breaks down the phytate in corn seed, thereby releasing phosphorus (and other minerals that phytate sequesters) in a form that pigs' bodies can use. But this approach also costs money and suffers because this enzyme sometimes denatures (breaks down) prematurely (e.g., if overheated during production or storage).

Is there another economical way to deliver usable phosphorus to swine, preferably in an environmentally acceptable manner? Some biotechnologists think the answer lies in transgenic "enviropigs," genetically engineered to produce phytase on their own. Genes for phytase could be isolated from bacteria or other microbes and attached to promoter regions specifically directing gene expression in mammalian salivary glands. The molecular construct would then be inserted by GM techniques into swine. If all went well, saliva in the transgenic hogs would soon be dripping with the precise enzyme needed to convert the natural phytate in ingested corn kernels to phosphorus compounds that swine could use. This would mean no more supplemental feeding of costly phosphorus or phytase to the pigs, no need to abandon otherwise nutritious corn kernels as hog feed, and less ecologically damaging phosphorus being excreted from the animals.

In 2001, scientists succeeded in engineering experimental pigs in almost the exact manner described above. These GM porcines produced copious saliva containing phytase from a bacterial transgene, and the enzyme appeared to work as intended in the pigs' mouths and guts, releasing usable phosphorus from phytate. As a result, the GM pigs required almost no dietary supplements of inorganic phosphate for normal growth, and, as an added environmental benefit, the animals excreted up to 75% less fecal phosphorus than non-GM controls.

Due to economic costs, social factors, and ecological considerations, these transgenic pigs would more likely be used in high-density hog ranches rather than on small family farms. Similar GM approaches can be envisioned for improving phosphorus management on massive chicken farms, another prime agricultural source of environmental pollution. On the other hand, some people argue that it might be better, in the long run, to rethink the entire concept of high-density ranching and high-tech genetic solutions (although the alternatives are not necessarily clear). They abhor the basic concept of mass-rearing "artificial" farm animals under artificial and perhaps inhumane conditions, especially when environments also are harmed by the process.

So, before transgenic enviropigs can have significant impact on ranching practices, several important hurdles remain to be cleared, including gaining acceptance by regulatory agencies and the public. This new approach to the phosphorus problem holds some promise, but its potential is far from realized.

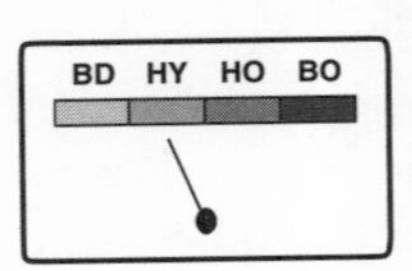

Mice as Basic Research Models

Mice are barnyard animals too, albeit not in the conventional meaning of the phrase. The mice discussed here call the laboratory home, rather than the farm, and their inclusion in this chapter is meant to highlight the underlying worth of basic or pure scientific research on GM animals. Laboratory mice (*Mus musculus*) have been at the forefront of the GM revolution, consistently yielding new insights on mammalian reproductive technologies and cellular operations.

Mice are favored for experimentation because of their small size, short generation time, large litter size, ease of husbandry, and the longstanding commercial availability of purebred strains for genetic analyses. In 2002, the complete sequence of the mouse genome was published, and this knowledge will further assist researchers who use mice as experimental models in genetic engineering. This essay describes key historical scientific breakthroughs on mice that will provide a backdrop for subsequent essays on GM methods and whole-animal cloning in livestock and other bona fide farm animals.

In principle, one way to engineer a transgenic mammal is to isolate a one-celled embryo (zygote) from a newly pregnant mother, fuse to it a desired piece of foreign DNA, put it back into the reproductive tract of an adult female, and hope for the best. In 1980, Jon Gordon and collaborators did precisely that. They microinjected short pieces of viral and bacterial DNA into one-celled mouse embryos, implanted the embryos into surrogate mothers, and later witnessed the birth of transgenic baby mice. Because the genetic transformations were initiated at the single-cell stage, after which the mitotic cell divisions occurred during embryonic development, each resulting mouse carried the transgene in all of its cells, germline as well as somatic. Thus, during normal sexual reproduction, the GM mice could transmit the transgene to their own natural offspring. Similar genetic engineering experiments with mice soon showed, also for the first time, that transgenes inserted by this method could alter important phenotypic features of GM mammals.

However, mammalian zygotes are tiny (each about one-tenth of a millimeter across), hidden deep inside the female, relatively few in number (compared to those of many fish, for example), and difficult to reimplant and coax to full term through a pregnancy. So, although this direct approach to genetic modification is conceptually straightforward, it can be difficult in practice. Accordingly, researchers also began to explore alternative GM routes, the two most successful of which proved to involve embryonic stem (ES) cells and cloning via nuclear transfer (NT).

Since 1885, it has been known that certain types of animal cells can grow and proliferate in artificial culture, such as a Petri dish, under suitable nutrient conditions. A century later, some special properties in this regard were discovered for ES cells in mice. Embryonic stem cells are generalized (undifferentiated) cells, taken from an embryo, that retain pluripotency—unbridled capacities to specialize later into new muscle cells, liver cells, heart cells, germline cells (eggs or sperm), and so on. In the early 1980s, researchers discovered that mouse ES cells properly cultivated in a Petri dish can be coaxed to retain this natural flexibility for a number of cell divisions. The pluripotency of ES cells, plus their relative ease of artificial culture, made these cells obvious targets for genetic engineering. The idea was to transform large numbers of cultured ES cells, for example by soaking them in foreign DNA, which the cells occasionally take up. Transgenic cells then would be returned to a multicellular natural embryo in a pregnant female. Researchers began accomplishing all of these feats with mice in the early 1980s.

Most of the resulting GM mice were chimeras; that is, they showed mixtures of two different genetic cell types: transgenic cells tracing back to the GM cell cultures and non-transgenic cells from the natural embryos. In effect, the live animals were only part GM. This also meant that the engineered mice transmitted their new genetic endowments to subsequent generations only when the transgenes happened to end up in cells destined to form an animal's eggs or sperm. This was not much of a practical problem in laboratory mice, which could be bred quickly, cheaply, and in abundance. But it could create difficulties for perpetuating GM dynasties in large, long-generation creatures like cattle.

For livestock engineering, this ES-cell method was less than ideal for another reason, too: Despite intense research, no one in the 1980s had succeeded in artificially culturing pluripotent or totipotent ES cells (those capable of directing the development of a full-term organism) from any mammal other than mice. The problems later were circumvented to some extent, but failures at the time prompted efforts to engineer strains of GM farm animals by yet another route, by cloning via nuclear transfer. Mice again paved the way.

In a 1981 paper, Illmensee and Hoppe announced the birth of three cloned mice by NT. In each case, the researchers extracted the nucleus from a totipotent embryonic cell by micropipette, transplanted it into an enucleated (nucleus removed) egg cell, allowed the newly generated cell to divide for several days in artificial culture, and then returned the early embryo (blastocyst) to the uterus of an adult female who later gave birth to a baby identical in genetic makeup to the original donor cell. At least, that's what the authors claimed. Despite repeated attempts over the next decade, other scientists were unable to replicate these experiments, and to this day nobody is quite sure why.

But in a historical sense, it hardly matters. The amazing claims in the 1981 paper spurred research efforts that led, two decades later, to the verified production not only of cloned mice by similar NT methods, but also to the routine production of cloned GM livestock (see following essays). Nowadays, elements of both the ES and the NT approaches often are used jointly when engineering GM farm animals. Embryonic stem cells are isolated from natural embryos, grown in artificial culture, and genetically transformed in a Petri dish. Nuclei from these transgenic cells are injected into enucleated eggs which then are reimplanted into the womb of a female. Later, if all goes well, a baby is born whose nuclear genome is a clone of the original transgenic cell.

As so often proves true, basic genetic research, in this case on mice, had unforeseen practical applications. As a way of perpetuating transgenic (or nontransgenic) barnyard animals with desirable genetic characteristics, whole-genome cloning, alone or in conjunction with other GM procedures, promises to alter the fundamental hereditary ground rules of farm-animal husbandry. The use of mice as model systems for genetic experimentation must be considered an unqualified success in generating important scientific insights for biotechnology.

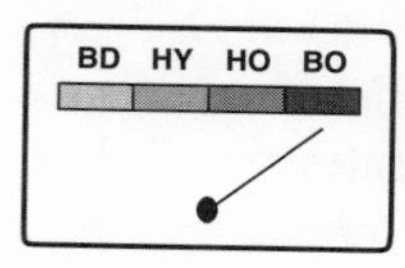

Hello Dolly

In 1997, a team of scientists in the United Kingdom introduced Dolly the sheep, the world's first mammal that had been artificially cloned from the cells of an adult. Genetically identical to her biological mother, lamb Dolly shocked the scientific world, not to mention the halls of philosophy, ethics, and religion. Many observers interpreted Dolly's birth as a blessed event coming from a welcome new technology that, if extended to create genetic copies of many top-notch farm animals, could result in far more productive ranches. Others saw the same event as a curse, a perilous first step down a slippery moral slope that someday might lead to the dreaded assembly-line production of human clones.

The techniques that Ian Wilmut and colleagues employed to produce Dolly are straightforward in outline. Using small needles and direct micromanipulation techniques, these researchers removed the nucleus (with its entire DNA contents) from a mammary cell of an adult Finn Dorset ewe, transferred it into an enucleated egg cell from a Scottish Blackface, and stimulated the egg, with electrical pulses, to respond as if it had been fertilized. This egg then began developing into an early embryo that was returned to the womb of the Blackface ewe. Several months later, the Blackface gave birth to Dolly, a genetic clone of her Finn Dorset genetic mom. (Strictly, the clonal identity between Dolly and her mother applies only to these animals' nuclear DNA; Dolly's mitochondrial DNA, housed in the cell's cytoplasm rather than nucleus, was inherited from the Blackface ewe.)

Similar nuclear-transplantation experiments using amphibians (as well as mice) had begun decades earlier. In the early 1950s, embryologists first transferred cell nuclei from frog embryos or tadpoles into enucleated frog eggs and thereby produced newly cloned embryos, each genetically identical to its sole parent. Some of these survived to the tadpole stage. In the 1970s, British researchers John Gurdon and colleagues extended the approach using nuclei transplanted from cells lining the intestine of adult frogs. Overall, the success rate in these experiments was inversely related to the age of the donor cells, being highest when the transplanted nuclei came from undifferentiated cells of frog embryos, but succeeding only rarely (<2% of attempts) when the nuclei transferred were from the cells of adults.

Despite such earlier research, the procreative feat that produced Dolly astonished almost everyone. Before then, biologists supposed that the genomes of specialized adult cells, such as those found in mammary tissue, normally had lost their ability to supervise the full-blown formation of any creature as complex as a mammal. In other words, they assumed that most differentiated cells, unlike early embryonic cells, were no longer totipotent, (i.e., capable of regenerating a new organism). They knew that during animal embryogenesis and growth, DNA in each cell type is altered (e.g., via chemical modification) in tissue-specific fashion, as particular ensembles of genes are activated or silenced. Although each adult cell still carries a full complement of DNA, the choreographed molecular ballet of normal development was thought to channelize or progressively constrain genomic potential. Thus, presumably the aged genome of each differentiated cell had lost forever its youthful vigor, its totipotency.

What Dolly suggested, however, is that a fountain of youth might exist, that the straitjacket of genomic aging is not inescapable. Apparently, under special circumstances the genomes of at least some differentiated adult cells retain sufficient flexibility to orchestrate the formation of an entire animal.

What were these special circumstances? In Dolly's case, the udder cells removed from her genetic mother first were grown in artificial culture under nutrient-poor conditions, forcing them to enter a quiescent state. In this resting or nongrowing stage of the cell cycle, most genes are inactive. This seemed to be a key factor in the genome's ability, when engineered into the refreshing intracellular environment of an egg cytoplasm, to reprogram its patterns of gene expression. Regardless of the precise explanation, the biotic engineering worked.

The media hoopla surrounding Dolly makes it easy to forget that the path to her creation was paved with failure. In the initial experiments using adult cells, nearly 300 nuclear transfers were attempted, but only Dolly survived the entire process to complete normal development. Still, it was a landmark technological achievement that, as we will see, raised tremendous hopes and soon spawned a mini-industry devoted to similar clonal propagation of several kinds of livestock plus other farm animals.

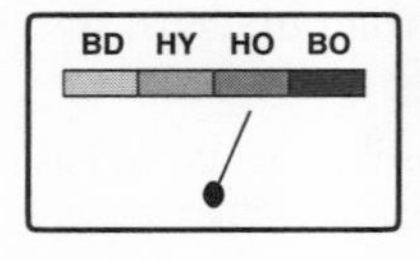

Cow Clones

Just six months after lamb Dolly pranced onto the world stage, calf Gene made his stunning debut as the first human-cloned bovine. Using variations on the laboratory method that produced Dolly the sheep, scientists at the upstart biotech firm Infigen engineered the infant Holstein bull using immature cells from Gene's unborn genetic father.

Here's how the process worked: From the reproductive tract of Gene's pregnant grandmother, a minuscule pre-embryo (Gene's sole parent) was removed. Cells in this diploid mass then were separated from one another, and, using a weak pulse of electricity, individually fused with enucleated bovine eggs. The latter were implanted into foster cows, most of whom aborted. But one calf, Gene, was born and survived. His genome was a perfect replica of the genome within the cells of the deceased intrauterine bull who, in effect, was Gene's biological sire as well as clone mate.

Actually, this technological feat was not as remarkable as that which produced Dolly, because the starting material in Gene's case was a young, undifferentiated cell, more likely to be totipotent. The achievement, nonetheless, was seminal because it showed that artificial cloning methods could be extended to other farm animals and could use other tissues. Since then, hundreds of experimental cows have been cloned from various bovine tissue sources.

Ironically, most of these carbon-copying efforts began with older or otherwise problematic parental cells. One 1998 study produced eight clonal calves from the somatic ovarian cells of an adult cow, and in a 2000 study six calves were successfully cloned from an adult cow's skin cells. These had been grown artificially (in Petri dish media) for up to three months. Later, eight calves were cloned using cells from an elderly cow who had become too old to reproduce by natural means. In 2002, the use of old bovine cells was taken to an extreme when scientists successfully cloned a calf, named K.C., from a *k*idney *c*ell of a dead adult who had been slaughtered for meat 48 hours earlier.

One reason for cloning from mature, as opposed to fetal, bovine cells is that many of the genetically influenced features that interest ranchers (such as uniform meat quality or high milk production) are only assessible in adult animals. For example, the cow mentioned above who posthumously donated kidney cells illustrates how bovine meat can be graded for quality after an animal's death, as is the norm in slaughterhouse operations, and yet still provide suitable genomic material for cloning live animals (like K.C.) that are genetically identical to the dead parent.

In some cases, prized cells to be cloned into new cows are themselves transgenic, having been engineered by earlier recombinant DNA methods. A case in point involves a GM Jersey cow named Annie, who was created as follows: First, cow cell nuclei were engineered to carry a bacterial transgene specifying lysostaphin, a protein offering protection against *Staphylococcus* bacteria that can cause mammary-gland infections known as mastitis. The nucleus from one of these transgenic cells was slipped into an enucleated bovine egg, which then was implanted into the womb of Annie's surrogate mother.

Months later, Annie was born, her cells genetically identical to the engineered construct that began the cloning process. At the time of this writing, Annie is not yet of age, but when she begins producing milk, she and her descendants should be far less susceptible to mastitis.

In other cases, prize bovines to be genetically cloned arose during the more normal course of ranching events. One remarkable example is Margo, a Holstein who single-handedly produced more than 55,000 pounds of milk when in her prime at two years of age. Margo passed away at age four, but some of her ovarian cells were collected postmortem and used as DNA donors for artificial cloning (as described earlier for K.C.). The result was Margo II, a genetic replica of Margo, who carried on her clone-mate mother's milk-producing tradition. Another famous cow was Mandy, a two-time All-American Holstein and three-time All-Canadian winner. Her cells were also gathered and cloned. In 2001, Mandy's clone-mate daughter, Mandy II, became famous in her own right as the first clone ever purchased at public auction. She sold for $82,000.

Some cloned cows have gone on to produce healthy calves of their own by conventional means. For example, two cloned Holsteins named Cookies and Cream, previously housed at the Minnesota Zoo in Apple Valley as goodwill ambassadors of their kind, recently gave natural birth to a heifer and a bull calf, respectively. Two other cloned Holstein heifers, Carbon and Copy, are now carrying on the zoo's public education effort on GM farm animals. Such cases suggest that cloned cattle can be healthy and normal, as supported by more formal scientific studies. Nonetheless, the success rate in cloning cattle usually remains low, with at least 85% of the embryos dying before or after transfer into their surrogate mother's womb.

Many of the cows cloned to date were given cute names (like Carbon and Copy), reflecting the fact that they are much in the public eye following directed news releases by self-promoting companies and universities. Undoubtedly, the biotech industry will soon begin to milk such cloned animals for all they are economically worth, but the extent to which this may benefit society remains to be determined.

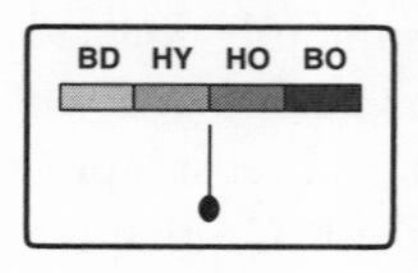

Barnyard Bioreactors

One year after lamb Dolly's birth, along came lamb Polly, in some ways an even more significant genetic engineering feat. Polly was not only a cloned farm animal, but also a transgenic one: The DNA from the fetal cell used to clone Polly first had been equipped with a human gene. Indeed, the cloning was merely a secondary exercise, meant to perpetuate a new biological home for the transgene that was of primary interest.

The transgene coded for factor IX, a blood-clotting protein useful in the therapeutic treatment of human hemophilia B. Blood plasma is the traditional pharmaceutical source of factor IX. However, the extraction procedure is expensive and risky because human plasma can contain disease agents such as viruses responsible for hepatitis and AIDS. If factor IX could be expressed at high levels in transgenic sheep, its recovery in commercial amounts would be much less expensive. The process should be safer, too, because during genetic transformation and cloning, a protein-coding gene is cleansed of contaminants from human tissues. Sure enough, Polly grew up to secrete high levels of human factor IX in her milk.

Polly was the first transgenic farm animal produced by artificial cloning methods, but she was by no means the first to carry and express a foreign gene of potential pharmaceutical importance. Nearly a decade earlier, scientists at the same research institute (Roslin, near Edinburgh) had engineered another ewe, Tracy, to carry a human gene for α_1–antitrypsin (AAT). They first attached the AAT gene to a sheep-DNA promoter region that enables protein production in mammary glands. With microsurgical instruments, the scientists then transferred this genetic construct into an ovine ovum (egg cell), grew the resulting embryo briefly *in vitro*, and transferred it to the womb of a foster ewe. Born in 1990, Tracy matured to secrete copious amounts of AAT in her milk. Several years later, AAT proteins from other GM sheep underwent clinical trials for the treatment of human lung disorders such as emphysema and cystic fibrosis. In the United States and Europe alone, more than 350,000 people suffer from serious lung ailments potentially amenable to amelioration by such AAT therapy, which limits the inflammation of lung cells.

Experimental GM approaches likewise have been used to harness a variety of farm animals as living bioreactors to produce human therapeutic proteins. Other examples of GM substances, currently under development or in clinical trial, are (1) in transgenic cows, human collagen for the treatment of burns and bone fractures, fertility hormones for contraceptive vaccines, lactoferrins for human gastrointestinal disorders, and albumins for surgery and trauma management; (2) in transgenic goats, antithrombin to prevent blood clots, TPA (tissue plasminogen activator) for treating heart attacks, and dozens of different monoclonal antibodies of potential use against human cancers; and (3) in transgenic sheep and pigs, factors VIII and IX for hemophilia, protein C as an anticoagulant, fibrinogen proteins for burns and surgical applications, plus human hemoglobin for blood transfusions.

Even rabbits have been engineered to produce therapeutic drugs. Pompe's disease is a rare and fatal genetic disorder that destroys muscles in infants, who are unable even to roll over or sit up. It results from the lack of an enzyme (α-glucosidase) that converts stored glycogen into glucose. In the late 1990s, transgenic rabbits were engineered that produce this enzyme and secrete it in their milk. When administered to humans in clinical trials, two infants with Pompe's disease showed vast improvement: One child later could sit up on her own, and another learned to ride a tricycle.

In several respects, GM "pharm" animals should be ideal sources of pharmaceuticals. First, they are large-bodied, so significant quantities of therapeutic human proteins can be extracted from their blood, milk, or urine. Second, the genomes of goats, pigs, cattle, and sheep generally function quite like those of people, meaning that human genes properly engineered into these farm mammals (more so than into microbes) are likely to yield usable therapeutic products. Finally, having had about 10,000 years of practice, humans already know how to raise barnyard animals efficiently.

Will the promise of windfall profits from pharmaceutical production revolutionize the family farm? Oddly, the answer is no, for the following reason: Calculations suggest that only small herds will be necessary to meet the global demand for most human therapeutic proteins. For example, based on milk yields per animal, milk protein contents, and purification efficiencies, as few as 300 transgenic cows or 8000 transgenic goats should satisfy the total annual global demand for human serum albumin. Via similar reasoning, a mere dozen transgenic sheep or a score of pigs could suffice for protein factor IX, and about 3 dozen cows or 1000 pigs for protein C. Although such transgenic herds could enrich pharmaceutical companies and be of huge significance to human health, they probably won't keep many ranchers busy (unless, of course, literally thousands of different pharmaceuticals are someday each engineered into a different stock of GM farm animals).

Nonetheless, entrepreneurial ranchers might wish to keep an eye on related genetic engineering possibilities. For example, most grocery-store milk comes naturally fortified with calcium, iron, phosphorus, and vitamin C from the cows, plus vitamins A and D added by human processors. If cows could be genetically engineered to produce dietary supplements such as vitamins A and D in ideal amounts, the marketing potential could be vast.

One such possibility under active consideration involves lysozyme, an antimicrobial protein found naturally in mammalian milk. Using appropriate GM technologies, lysozyme production potentially could be increased by more than 1000-fold in engineered cows. If such milk did not harm the cows, if it conferred significant health benefits to people who drink it, if it passed regulatory muster for safety, and if consumer acceptance was high (several big ifs), then GM cows could have a broad impact on farming practices.

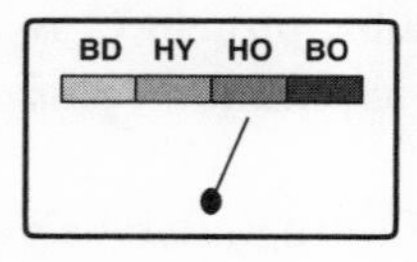

Vaccinating for Animal Health

Genetic engineers have harnessed farm animals as living factories to churn out various pharmaceuticals for human health. Have they also used their engineering wizardry to improve health in the animals themselves? To human sensibilities, many farm animals reared for production lead appalling lives. On the other hand, nobody benefits from malnourished or sick animals, so economic as well as ethical considerations should provide incentives to engineer genes for better health into barnyard animals.

One example already mentioned involves mastitis in cows (see "Cow Clones"), where scientists recently cloned a heifer carrying a transgene for lysostaphin, a protein that thwarts bacteria that cause this inflammatory disease of the udder. Someday, this kind of genetic engineering might be extended beyond the experimental stage to offer real protection against mastitis and a host of other mammalian diseases.

In the fight against disease agents in domestic animals, another intriguing GM approach, "DNA vaccination," was prompted by an article that appeared in a 1990 issue of the journal *Science*. Researchers demonstrated that pieces of viral DNA, when engineered into a suitable expression vector and injected into mice, were incorporated and successfully activated in the animals' muscle tissue, thereby potentially inducing a protective immunity against the virus. Ten years later, another research team showed that such immunity potential could be realized. In this case, they inoculated rainbow trout with a plasmid DNA concoction containing an antibody sequence specific for a rhabdovirus that causes fish hemorrhagic septicemia (a debilitating blood infection). The fish responded by producing antibodies that afforded protective immunity against the virus.

Such recombinant DNA vaccines include specific transgenes that, after injection, activate within the recipient's body to produce and distribute antibodies against targeted disease agents. Thus, they differ from the whole-virus potions that are the traditional vaccines used in animal (and human) medicine. In these conventional vaccines, the injected virus is first debilitated or killed, for obvious reasons. However, the disabled virus still presents health hazards (they are somewhat pathogenic), and they may be less than fully effective in eliciting immune responses. The new DNA vaccines, it is hoped, can avoid health risks of traditional vaccines, while still inducing protective immunity.

DNA vaccines have recently been used in a few commercial fish-farming operations where diseases are of special concern due to high animal densities and unsanitary rearing conditions. For example, in some salmon farms (see chapter 6), particular specimens are physically injected with DNA vaccines in efforts to control the spread of infectious diseases. Research suggests that even a modest fraction of protected individuals can significantly inhibit major disease outbreaks.

However, DNA vaccines are not yet used extensively in standard veterinary medicine, for at least two reasons. First, their titers tend to be low, so the need for high dosages and repeated administration make their costs prohibitively high. In the future, adjuvant molecules that improve immune responses might be used to circumvent this difficulty. Second, pathogen variability is a potential drawback that could compromise the use of DNA sequences that are highly specific in antibody elicitation. This problem might be overcome by injecting cocktails of DNA sequences with different antibody specificities. These and other possibilities likely will be explored in upcoming rounds of research on genetically engineered DNA vaccines. The findings may also uncover new routes to protection against human pathogens that are refractory to traditional vaccines.

Annual sales of veterinary biologicals total about $2.5 billion in the United States and Europe alone, and large markets also exist in Asia and elsewhere.

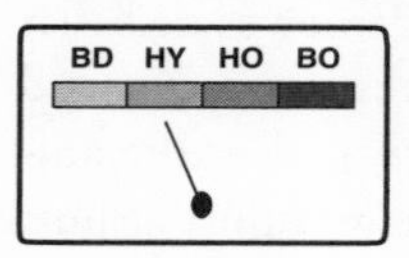

Conventional vaccines have had a huge impact in the war against diseases in domestic animals, but more and better protection is desirable. It remains to be seen whether, and to what extent, recombinant DNA vaccines will help.

Engineering Foods for Animals

Another route to better animal health is through improved feed and nutrition. Much of traditional agriculture is devoted to delivering suitable corn, hay, and other fodder to barnyard animals, but high-tech industries have come to the table in recent years to improve the plant-food products or to add dietary supplements to animal feed. Such efforts are laudatory in principle, but they can sometimes backfire with dire consequences, as evidenced by an ongoing debacle in antibiotic use.

Livestock producers in the United States place an estimated 24 million pounds of antimicrobial drugs in animal feed annually. The aquaculture (fish and shellfish) industries use another 300,000 pounds. These antibiotics often are administered in subtherapeutic doses to improve animal growth (by as much as about 10%), but an emerging development—microbial antibiotic resistance—is causing great alarm among scientists and healthcare professionals.

As noted previously, microbial species tend to evolve genetic resistance to the antibiotic compounds that humans use most widely to kill them. Typically, resistant strains begin to emerge within a decade or less of antibiotic deployment. For example, tetracycline was introduced in 1948, and microbes resistant to it first appeared in 1953. For several other major antibiotics, historical dates in the drug-discovery/drug-resistance cycles are: penicillin, 1928/1946; sulfonamides, 1930s/1940s; streptomycin, 1943/1959; chloramphenicol, 1947/1959; erythromycin, 1952/1988; vancomycin, 1956/1990; methicillin, 1960/1961; and ampicillin, 1961/1973. In recent years, the development of microbial resistance has outpaced the rate of drug discovery, seriously threatening an end to the era when antibiotics can be called upon to treat major outbreaks of human (and animal) diseases.

Much of this microbial resistance probably originates in response to antibiotics improperly deployed in human medicine, but there is growing concern that drugs added to animal feed tremendously exacerbate the problem. About 40% of all antibiotics manufactured in the United States go into farming and ranching uses, and they include such drugs as penicillin, tetracycline, and fluoroquinolones that have been hugely important in human medicine. The dangers are so great that the European Union, Japan, Australia, and New Zealand, among others, have banned the subtherapeutic use of such compounds in animal feed. The United States and Canada are among the few major countries yet to see the light, having caved in to agribusiness and farm-state interests that have worked vigorously to block much-needed legislation in these areas.

Although certainly high-tech, the use of antibiotics in animal feed is not really a part of the genetic engineering revolution because recombinant DNAs are not centrally involved. Nonetheless, the lessons for genetic engineering are serious; namely, because animal feeds are ubiquitous, economically valuable, and potentially interwoven with human welfare, great care must be exercised in any contemplated scheme to improve them genetically. Of course, these same points can be used to suggest great possibilities for the genetic improvement of animal feed.

One recent example suffices to illustrate some of the experimental approaches and objectives of ongoing research on GM animal feed. Malt or barley (*Hordeum vulgare*) seed might be an economically desirable alternative to corn and soybean as chicken feed, were it not for its low nutritional value for birds. This stems from a poorly digestible constituent of barley seeds, β-glucan. Unable to break down this polysaccharide (a type of carbohydrate), chicken intestines extract only limited nutrients from barley seed and also produce sticky droppings that adhere to the birds and their cages. To address this problem, scientists engineered a transgenic malt containing glucanase, an enzyme that processes β-glucan. When added in small amounts to normal barley and fed to chickens, the birds showed significant weight gain and excreted less sticky droppings.

As this type of GM research is extended to additional animal feed crops, four questions should be routinely addressed: How do the genetic alterations affect the growth and performance of the plants themselves? How do they affect the health and performance of the fed animals? Do they have any carryover consequences with regard to human health? And how do they impact the broader ecology of the farmlands in which these GM products are grown and utilized? Direct economic incentives normally will drive biotech industries to answer the first two questions themselves. Answers to the latter two, however, often will require regulatory oversight or government intervention to ensure that broader societal interests are served as well.

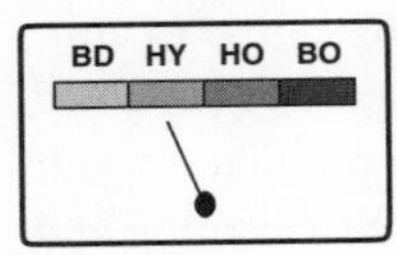

To date, GM foods for domestic animals have received considerable discussion, but limited action. Before the overall evaluation can be moved from the hyperbole category, more will have to be accomplished.

Cloned Organ-Donor Pigs

For a while, pigs proved to be harder to clone than mice, sheep, goats, and cows. After a decade of glitches and mostly failed attempts, the first accounts of successful pig cloning by nuclear transfer appeared in 2000. Since then, the cloning of pigs has advanced toward two ultimate research goals, one predictable and the other a bit more surprising.

The more humdrum goal is to produce genetic carbon copies of prize pigs carrying desirable genes (e.g., for outstanding meat production or better environmental performance) that either arose naturally or were engineered using transgenic techniques. As an example, in February 2002 Infigen Inc. announced that it had used NT methods to clone piglets from two elite non-GM boars: a prize winner by the name of "401K" that a semen company had purchased for $43,000; and "The Man," a champion Yorkshire that commanded $77,000.

The less obvious goal of pig cloning is to assist xenotransplantation efforts. Over the years, hundreds of human patients have been exposed to live pig cells and tissues (notably heart valves) in experimental therapies for disorders of the liver, spleen, and blood. But the revolutionary hope for the future is in the use of pig organs in human transplantation surgery. The pig is a promising animal for organ donations because it is similar in size and physiology to humans and is abundant in supply compared to organ-donor humans or other primate species that otherwise might be preferable. However, xenotransplantation research long foundered because of concerns that the human immune system would reject xenotransplanted organs and that pig viruses might be transferred into our species as new disease agents. Geneticists have shed some recent experimental light on both of these topics.

With regard to the immunological issue, one pig enzyme that acts as a key antigen in eliciting hyperacute rejection by human recipients is $\alpha_{1,3}$-galactosyltransferase (GGTA1). In 2002, two companies simultaneously reported the production of cloned "knockout" piglets (see "Knockouts and Resuscitations" in the appendix) engineered to be genetically deficient for this enzyme. The GGTA1 gene was targeted and inactivated in fetal pig fibroblast cells grown in culture, and the nuclei from these GM cells were microtransferred into enucleated pig oocytes. Then, after gestation in foster-sow wombs, the cloned piglets were born. Although a major technological achievement, this experiment is merely one preliminary step toward overcoming the organ-rejection hurdle. For example, several other proteins in pigs likewise elicit antigenic responses in human tissues, and likewise will have to be disabled genetically (or their antigenicity otherwise circumvented) before cloned GM pigs can become suitable organ donors.

With regard to the transfer of viruses, all pigs contain copies of porcine endogenous retroviruses (PERVs) that some doctors fear might hitch rides on donated organs and result in malignancies or other disorders in human recipients. This is also a public health matter because once in humans, PERVs conceivably might transfer from person to person. In initial experiments using mice as a model, researchers examined whether transplanted cells from a pig's pancreas could transmit PERVs to another mammalian species. Indeed they could, a finding that raises a huge cautionary flag against extending such transplantations into humans.

There certainly are precedents for newly emergent diseases crossing from animals to humans with deadly effect. Recent examples included "mad cow disease" transmitted from cattle to humans, avian influenza virus that has killed several people exposed to infected birds, and the AIDS virus which probably originated in other primate species and jumped to people. Nonetheless, proponents of pigs as organ donors point out that among the numerous patients that already received pig blood plasma or tissue extracts for various therapeutic purposes, no instance of human PERV infection has surfaced.

In the United States at any time, more than 60,000 desperately ill people await surgical transplantation of one organ type or another; a new name is added to the list every 18 minutes. The limiting bottleneck, in helping these people, is the small number of suitable organs available from human donors. In 2000, one of the first cloned piglets was named Xena, perhaps in reference to the hopes for her kind in future xenotransplantation efforts that would overcome the shortage of organs for transplantation. In 1995, a pioneering company in xenotransplantation research predicted that animal organs would be available for human use by 2002. This date has come and gone, and it now appears that at least several more years may pass before pig-organ xenotransplantation might reach any substantial clinical use.

Furthermore, tissues engineered from human stem cells (chapter 7) and artificial organs may well supplant cloned pig parts as the primary foci of future medical efforts. Especially for this reason, and until proven otherwise, pig-organ xenotransplantation must for now receive an evaluation in the hyperbole range.

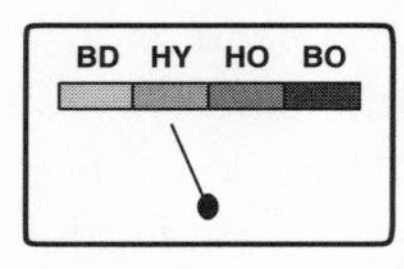

Possibilities with Poultry

Mammals are not the only farm animals being tinkered with genetically. Chickens are likewise the experimental subjects of genetic manipulation for trait improvement, better breed performance, or as living bioreactors for pharmaceutical proteins. Growing numbers of companies (including Origen Therapeutics of Burlingame, California, Viragen of Plantation, Florida, AviGenics of Athens, Georgia, TransXenoGen of Shrewsbury, Massachusetts, GeneWorks of Ann Arbor, Michigan, Vivalis in France, and others in South Korea and China) now focus primarily or exclusively on engineering transgenic poultry.

As targets for genetic manipulation, chickens offer several research challenges, as well as opportunities compared to mammals. Normal cloning procedures are impractical because avian oocytes cannot be removed from the bird, altered, and reimplanted as can those of cows, goats, pigs, or sheep. On the other hand, chicken eggs are laid in abundance (up to 330 per year by a modern, genetically selected White Leghorn), are cheap and easy to handle before hatching, and come with a large mass of sterile egg-white (3.5 grams of protein per egg) that serves as a favored site for the incubation of transgenic biopharmaceuticals. Furthermore, foreign proteins expressed in this proteinaceous mass can be harvested immediately from eggs, whereas those from the milk of transgenic farm animals must await the onset of lactation.

The nature of the avian egg is key to appreciating both the challenges and the recent progress in poultry genetic engineering. A hen ovulates once per day, and if she has been artificially or naturally inseminated, the oocyte is fertilized immediately in the innermost reaches of her oviduct (the female reproductive tube). Over the next 22 hours, the egg travels slowly down the oviduct, where albumin (egg white) is deposited around the yolk and embryo, followed by the protective egg-shell membranes and then the hard shell itself. Cell division is rapid during this time, such that the tiny embryo, lying atop the yolk, consists of about 60,000 mostly undifferentiated cells by the time the egg is laid. While inside the hen's body, the embryo is not easily procured for genetic manipulation, but once the egg is laid, quick and harmless access can be gained by injecting it with a fine needle, or, if need be, by cutting a small window in the shell that later can be resealed.

Transgenes have been delivered successfully to chicken embryos by microinjection and via GM viruses or transposable elements (see appendix). One ticklish physical approach is to deposit cultured chicken cells containing foreign DNA either into the yolk area near the embryo shortly after ovulation, or later underneath the embryo in a freshly laid egg. A more efficient delivery method is by retroviruses. In early experiments, however, the virus itself also proliferated in the birds, making the approach less than ideal. Subsequent improvements led to a replication-deficient virus as the delivery vehicle and coupling the entire procedure with a rapid screening system to detect any chicks that might be transgenic.

Because of the multicellular nature of the avian embryo at the usual time of genetic manipulation, what often emerges is a chimeric chicken. In other words, a hatched chick typically consists of many cells that carry the transgene, but others that do not. This can be a problem, especially when establishing transgenic chicken lines, because the luck of the draw pretty much determines whether a transgene in a chimeric bird ends up in gametes that will produce the chicks of the following generation.

Undaunted, poultry-engineering firms are flying ahead with grandiose plans to improve domestic fowl. Some of these companies are tight-lipped about what they are up to, however, even by the normal secretive standards of the biotechnology industry. GeneWorks, for example, has refused to publish its findings in peer-reviewed scientific journals, for competitive and proprietary reasons. Patent-right issues produced at least one court battle involving two competing firms (AviGenics and Viragen). The net result is that precious little has been published on the actual applications of chicken engineering, and the examples that follow were gleaned from (questionably reliable) press releases and company websites.

Pharmaceutical proteins from laid eggs are the primary goal of most ongoing research into GM chickens. In the pipeline are transgenic birds whose egg whites someday might deliver commercially viable quantities of cancer-treating drugs and other proteins of potential value in human medicine. Because of their low cost and high productivity, chickens (more so than livestock) may be especially useful for making high-dose, low-potency drugs like serum albumin and various antibiotics. For delivering such therapeutics, chicken eggs hold another advantage over mammalian milk. Any human protein expressed in the milk of a transgenic mammal is likely to be similar in structure to the animal's own counterpart protein, making purification of the human product difficult or expensive. Chicken eggs may be less problematic in this regard because the bird's endogenous proteins are somewhat more distinctive from those of humans.

Other goals of poultry engineering are to improve the value or health of the chickens themselves. Companies envision the generation of GM chickens with improved disease resistance, faster growth rates, less fragile bones, or a bigger breast, for example. Another notion is to alter the proportions of white meat (generally preferred in many Western countries) and dark meat (favored in Asia) in various GM breeds to better suit local markets.

Animal-rights activists and others are concerned that the economic value and the well-being of a GM chicken sometimes oppose each other. A case in point was the recent announcement in Israel of the creation of naked GM chickens. The preplucked birds lack feathers, making them easier to process at the broiler factory, but at what cost to the animal's health and comfort before death?

The GM poultry industry's history of secrecy on hard scientific issues, contrasted with its grandiose claims yet limited published accomplishments to date, necessitate that I give this enterprise only a hyperbole ranking at present. I hope this grade will improve in the future.

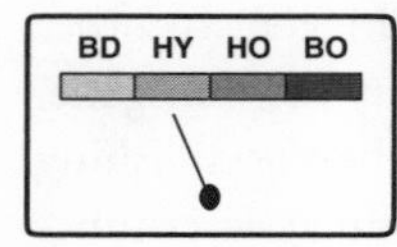

Copy Cats

You might suppose that breaking the technological cloning barrier for sheep or cows would quickly have opened similar capabilities for all mammals. But this has not always been the case in practice; some species have proved much easier to clone than others. Sometimes, poor knowledge concerning the basic reproductive anatomy or physiology of a species has been the hindrance. Other times, the reproductive biology is understood reasonably well, yet it remains refractory to successful cloning manipulations.

Dogs (*Canis familiaris*) illustrate the latter situation. Compared to many other mammals, bitches come into heat infrequently and unpredictably. Also, the eggs they release from their ovaries are very immature and difficult to ripen in a test tube. Thus, getting viable ova has been a major obstacle in efforts to clone man's best friend. Undoubtedly, however, it is only a matter of time and effort before success in this endeavor becomes routine.

Successful cloning came a bit faster for the cat, *Felis domesticus*. In February 2002, scientists funded by GSC (Genetics Savings and Clone Company of College Station, Texas) reported the successful cesarean birth of the world's first cloned kitten. Named "CC" for Carbon Copy or Cloned Cat, this kitty's nuclear genome is identical to that in the cells of her calico progenitor. Carbon Copy is the first clone created in Project Missyplicity, funded at more than $3.5 million by an octogenarian Arizona financier, John Sperling. The word "Missy" is the name of Mr. Sperling's pet, a husky-border collie that Mr. Sperling would like to replace, as closely as possible, when the dog's time is up.

In generating the carbon-copy feline, the GSC scientists first tried to transfer the nuclei from adult skin fibroblast cells into enucleated cat eggs. Eighty-two cloned embryos were implanted into the wombs of surrogate mothers, but only one pregnancy resulted and that fetus died. Next, the researchers used the nuclei from two cat cells lining the oral cavity, and three from cumulus cells surrounding the ova. This time, one of the cumulus nuclei, from a donor cat named Rainbow, led to a viable kitten fetus and the eventual birth of Rainbow's younger clone mate, CC. Overall, then, the cloning success rate was 1.1% (1 in 87), roughly comparable to early figures for livestock during the first few years of animal cloning by nuclear transfer.

Critics have blasted the GSC program for attempting to create more companion animals in a time when untold numbers of pets are destroyed or abandoned each year for lack of suitable owners. Pet overpopulation is a serious problem. On its website, the GSC responds that in 2001 alone it paid more than $250,000 to spay clinics for eggs used in the experimental cloning and that the clinics can use that money in turn to sterilize far more cats than likely will ever be cloned (especially at the current estimated cost of $20,000 per cloned kitten). Furthermore, as Mr. Sperling illustrates, many people do not want just any cat or dog replacement for their current pet. When the time comes, they want a genetic copy of their specifically loved one.

In anticipation of the quasi-routine cloning of cats, dogs, rabbits, and other pet species in the foreseeable future, or at least to cash in on that hope, various companies already have started pet "gene banking." For an initial service charge (ca. $700) and a monthly storage fee (ca. $10), a company will store for posterity a collection of clonable cells from your favorite living pet. All you need do is visit your local veterinarian for a small skin biopsy from your animal companion and mail the tissue sample to the biotech firm. The company will grow the cells in culture to get them proliferating properly and then freeze (cryopreserve) the cells in perpetuity for possible future cloning. If cloning technologies do indeed come online, after your current pet has died, the company may be able to resurrect a genetic copy of him or her using nuclei from the stored cells.

In the United States alone, there are more than 50 million dogs and an even slightly greater number of cats. These range from the abandoned and pathetically homeless to some of the most loved and pampered creatures on earth. The latter may be lavished with exclusive accommodations, gourmet meals, exercise and training staffs, coiffure shops, and even tombstone cemeteries in which to spend their afterlife. Now it appears that some of these pampered pets might be granted a genetic afterlife as well, lived through their clonal descendants. It used to be said that cats have nine lives, but that number may have to be revised upward in the brave new world of pet cloning that some people advocate, but have not yet brought to wide fruition.

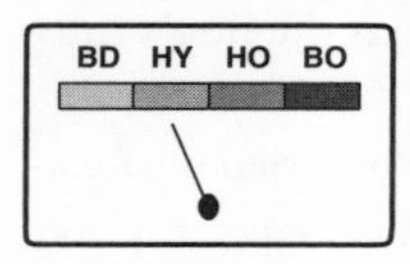

Good-bye Dolly

Whole-animal cloning offers the prospect of perpetuating carbon copies of a given genotype across generations. Does this imply that a newly cloned lamb or puppy will always be as good as the original animal that donated a somatic cell to the nuclear-transfer procedure? Perhaps not. The behavior, personality, and morphology of any complex animal is a product of more than its genes alone. The same dog raised in different environments could turn out friendly and playful if nurtured in a loving home but defensive or vicious if reared abusively. Similar themes apply to human beings, of course, and this should be a sobering reminder to anyone wishing to clone members of our own species (see chapter 7).

But there is another, genetic sense, in which a cloned animal might be diminished from the original, losing a bit of luster in the duplication process, much as a photocopy can fail to capture the full intensity of its template. This issue has arisen for Dolly the sheep and other cloned barnyard animals, especially those that have been engineered by the transfer of nuclear DNA from an adult cell. When inserted into an enucleated oocyte, is that "adult" DNA fully rejuvenated to the state of DNA in a freshly fertilized egg formed in the conventional way (i.e., by a union of egg and sperm)?

There are reasons to suspect it is not. Cloning procedures for barnyard animals typically yield far more failures than successes. The most common fate of a cloned embryo or fetus has been intrauterine death, and even on exceptional occasions when successful births occurs, health problems often emerge later on. Abnormalities commonly reported in fetuses and neonates include respiratory and cardiovascular irregularities, edema (excess fluids in tissues or in body cavities), defects of the urogenital tract, and severe deficiencies of the lymphatic system. Some of the problems might be attributable to poor NT technique or to spontaneous abortions identical to those in natural pregnancies, where early-life deaths are common also. However, another explanation also has been advanced: Perhaps some of the problems are symptomatic of incomplete genetic reprogramming.

There are several potential sources of reprogramming difficulties in animals cloned by NT because the procedure in effect bypasses several of the early genetic steps of normal organismal development. For example, methylations (attachment of methyl groups to particular nucleotide sites), acetylations (attachment of acetyl groups to histone proteins), and other changes in chromosome structure routinely accompany gametic and zygotic development during natural reproduction but probably do not occur in the same way during the process of NT cloning.

Another example involves sex chromosomes. In female mammals, both X chromosomes initially are active in normal zygotes, but an NT-cloned embryo receives (from a somatic donor cell) only one active copy of the X. This is because of a natural process in mammals, known as the Lyon effect, in which one of the two X chromosomes in each somatic cell of a female is inactivated during her development. Abnormal expression patterns in X-linked genes have been reported in the tissues of cloned fetuses, and this may contribute to the high frequency of spontaneous abortions observed in such pregnancies.

One specific kind of change in chromosome structure, the shrinkage of telomeres, has attracted special attention. Telomeres are the tips or ends of chromosomes, and they shorten as an inherent part of the natural aging process. During each generation of regular sexual reproduction, telomeres are rejuvenated (restored to their youthful length), but questions have arisen as to whether this occurs also during artificial NT cloning. If it does not, a cloned embryo in effect could begin life prematurely old. Indeed, short telomeres have been noted in several NT-cloned animals compared to age-matched controls produced sexually. However, such reports have been challenged in the literature on the grounds that telomeric lengths are hard to quantify and interpret. More intense scientific inquiry into this issue is needed.

Whatever genetic factors might harm the health of or cause premature aging in cloned animals, they would appear not to be universal or overriding. Many cloned barnyard animals have survived for years and appear to lead reasonably normal lives. What about Dolly the sheep, the first barnyard clone? She was generally healthy, and indeed produced healthy lambs of her own, the old-fashioned way. However, in 2002, at the relatively young age of five and one-half years (sheep can live for more than a decade), Dolly developed arthritis prematurely, and this raised some suspicious eyebrows. Less than a year later, in early 2003, she showed signs of progressive lung disease, and, sadly, had to be euthanized. Dolly had survived for only about one-half of a sheep's maximum life expectancy, but whether her death was mere ill fortune from a routine lung infection or relatable more directly to cloning per se, remains uncertain.

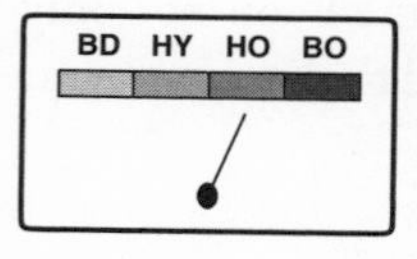

Concluding Thoughts

Barnyard animals have been genetically engineered (made transgenic and/or cloned) under four primary rationales: to provide model systems for basic inquiry into fundamental cellular operations and GM technologies; as living factories to biomanufacture therapeutic medicinal compounds; to improve animal performance (e.g., with regard to providing better food products, services, companionship, or less environmental damage); and to use as suitable organ donors for human transplantation surgery. The first two arenas already have delivered significant and tangible benefits, whereas the last two remain mostly in the early or intermediate stages of research and development.

If for no other reason than physical scale, most of these contemplated applications should pose no overt threats to the environment. For example, even small herds of suitable GM goats or cattle can be sufficient to satisfy the entire global demand for many of the pharmaceutical drugs to be derived from transgenic animals, and it is hard to imagine that pets or livestock will ever be cloned in numbers remotely approaching those generated naturally.

Thus, the situation in the GM barnyard can differ markedly from that in GM plant agriculture, where the engineered crops often must be sowed widely to achieve desired goals (such as increased yield via tolerance to herbicides or pests). In other words, GM farm animals and their transgenes in most cases can be kept on far tighter ecological leashes than can GM crops. This is additionally true because, unlike many crops, few barnyard animals are prone to come into contact and hybridize with related wild species.

With regard to human health, the issues raised by GM farm animals and GM crops usually are similar. In both cases, any foods or pharmaceutical products derived from the transgenic organisms should be assurity tested for absence of toxins, allergens, disease vectors, or other medical hazards before they are released to the public. The contemplated use of pig organs in surgical xenotransplantation offers a special medical screening challenge, because contact with human tissues would in that case be highly intimate and pervasive, and the consequences of introducing any new disease agent to our species could be dire.

Overall, genetic engineering in the barnyard is off to a fanfare start. In the coming years, it will be interesting to monitor whether the field matures and flourishes, or fizzles. If the former outcome is more nearly correct, as I suspect, then the broader enterprise could become a genuine boon not only for venture capitalists, but also for consumers, medicine, and the environment.

Fields, Forests, and Streams

Not content to tinker only with traditional agricultural crops and domestic animals, bioengineers are venturing farther afield to genetically alter a variety of plants and animals from forests, meadows, and aquatic habitats. Some of these GMOs blur the distinction between cultivated and uncultivated, or between domesticated and wild, but these transgenic creatures and organismal clones, however categorized, represent further evidence of the wide range of taxa, as well as traits, that can be genetically manipulated using recombinant DNA methods.

The extension of genetic engineering beyond factories, croplands, and barnyards raises many additional hopes and fears, especially with regard to the environment. Might genetic engineering experiments "in the wild" invite ecological disasters? Or, at the other possible extreme, could some of the new genetic discoveries and GM products actually help the planet to recover from insults of earlier technological endeavors (such as energy extraction, mining, farming, and commercial fishing) that often generated toxic wastes or otherwise damaged natural ecosystems?

Most of the contemplated GM projects relating to the environment are only in the early phases of research and development, so definitive answers to such questions will take some time to emerge from direct experience. In the meantime, given the general dearth of critical scientific evidence in this neophyte field, plausible arguments pro and con can be made regarding many prospects for outdoor genetic engineering. As is typical of GM technologies applied to nature, simple solutions can be elusive, and case-by-case analyses are required.

Pulp Nonfiction

Trees provide greater scientific challenges for bioengineering than do most herbaceous plants. First, the genomes of most tree species are poorly characterized at present, leaving open many basic questions about their gene structures and functions. Second, relatively few trees proliferate clonally in nature (notable exceptions being aspens and poplars from the genus *Populus*), so even if an individual tree is engineered successfully, subsequent propagation of its transgenes remains problematic. Third, trees usually take at least several years to mature, making it harder to monitor the multigenerational effects of any genetic tinkering.

Undaunted, some forest scientists are forging ahead with plans to redesign trees for a variety of genetic features of potential commercial significance. By the year 2002, the Animal and Plant Health Inspection Service (APHIS) of the USDA had received (and often approved) nearly 150 applications for outdoor field tests of GM trees, while dozens more such experiments are being conducted in at least 15 other countries. Many of these are private-industry ventures, proprietary, and rather secret. At the APHIS website, for example, the phrase "confidential business information" is all that commonly appears in the column listing the gene and the species of tree being studied. Other preliminary GMO ventures have received wider publicity.

A case in point involves attempts to genetically alter the lignin content of pulpwood trees. Lignin is a hydrocarbon substance that helps give tree trunks their stiffness, and industries must go to great lengths to remove it during the conversion of wood to pulp and paper. During the process, paper mills use huge amounts of highly toxic chemicals that are a serious source of environmental pollution. A 1988 study from North Carolina State University estimated that even a 5% reduction in the lignin content of pulpwood trees would appreciably lessen the need for these harsh chemicals and also save the paper industry about $100 million annually in processing costs. Clearly, the economic and environmental incentives to engineer pulpwood trees for reduced lignin content are considerable.

Toward that end, researchers produced a transgenic strain of aspen trees (*Populus tremuloides*) with 45% less lignin than their wild brethren. They manipulated genes in a biosynthetic pathway in such a way as to partially block the trees' natural lignin-making capacity. Remarkably, the GM trees remained sturdy and actually grew faster than normal. The structural integrity of the trees seemed to be maintained in part by a compensatory increase in cellulose, another carbohydrate that contributes to the natural tensile strength of aspens' cell walls. Similar attempts now are underway to genetically engineer reduced lignin content in pine and eucalyptus trees, two traditionally popular sources of pulpwood.

Critics point out potential drawbacks to genetically engineered trees such as these, especially if they someday were to be planted on commercial scales. Trees live far longer than most biotech food crops, so any negative long-term ecological impacts will be harder to assess and counter. How will soil-dwelling fungi and bacteria, not to mention beneficial insects and birds, respond to these structurally modified trees? Some of the ecological effects might be fairly subtle, but nonetheless important. For example, if low-lignin trees degrade faster after death, might this negatively impact species ranging from beetles to cavity-nesting birds who depend on the availability of slow-rotting wood?

Additional ecological questions arise: With such genetically altered features, how will the trees themselves fare in nature? For example, might low-lignin trees be more susceptible to insect attacks? Might the transgenes transfer via hybridization to other forest species and genetically modify them in undesirable ways? More generally, might not the introduction of GM trees further promote monocultural tree plantations that are little more than biological deserts? Already, vast uniform stands of nontransgenic pulpwood pines, planted, for example, in coastal regions of the eastern United States, are biotically impoverished, containing few other forms of life.

On the other hand, in a recent analysis of transgenic poplars grown experimentally for four years at sites in France and England, researchers concluded that lignin-altered trees remained healthy and interacted normally with insects and soil microbes. These GM trees also grew at standard rates, required smaller amounts of pulp-processing chemicals, and still yielded high-quality pulp.

Genetic engineers argue that in addition to reducing papermill use of caustic chemicals, GM trees can provide other environmental benefits. They suggest, for example, that forests engineered for higher productivity will enhance the efficiency of wood-product industries and thereby lessen harvesting pressures on native stands. Globally, timber products support an annual $400 billion industry, with demand for pulp and paper alone estimated to increase by about 50% in the next two decades. Some analysts contend that bioengineered trees offer the best way to meet growing demands for wood products without further decimating biodiverse natural woodlands.

For economic and biological reasons, forest genetic engineering has not yet proceeded in the frenzied, profit-driven atmosphere that accompanied the promotion of faster growing GM crops. This may be a blessing. Without the overwhelming pressure and capacity to make a fast dollar from slow-growing trees, societies and the sciences they support will have more time to process wise judgments about how GM pulpwood forests might affect both the economy and the environment.

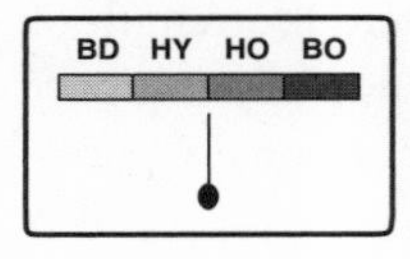

Antimalarial Mosquitoes

Mosquitoes are not just irritating to humans and other animals, they are also among nature's most dangerous vectors of disease. When mosquitoes suck their blood meals, they sometimes pick up and transmit microbes that are the causal agents of a wide variety of serious illnesses. A classic example is human malaria, which sickens 200–300 million people and causes about 2 million deaths annually, mostly among small children in Africa.

The microbial agents of malaria are unicellular protozoans from the genus *Plasmodium*. They have a complex life cycle. Haploid sporozoites, the parasite's invasive life stage, inhabit salivary glands of infected mosquitoes and are injected into a host animal when the insect feeds. The sporozoites then travel to the host's liver, where they proliferate asexually and are released into the bloodstream as merozoites. There, they multiply further and eventually cause red blood cells to rupture. The symptoms of malaria are due to anemia from blood cell loss and debilitating fevers from the host's immune response. Some merozoites may be picked up when another mosquito bites a diseased animal. In the mosquito's midgut, the parasitic cells fuse into diploid cells, which then undergo meiosis to form the parasite's next generation of haploid sporozoites. These migrate to the mosquito's salivary glands, thereby completing one full round of the *Plasmodium* life cycle. Different species of mosquito (e.g., in the genera *Anopheles* and *Aedes*), house and transmit different species of *Plasmodium*, which tend to be taxon-specific disease agents. Human malaria, for example, is due to *P. falciparum*, whereas an agent of avian malaria is *P. gallinaceum*.

As is generally true of host–parasite systems, infected humans or other animals are not just passive bystanders to the *Plasmodium* attacks. For example, some people have a form of the *HLA-B53* gene that provides partial protection against malaria by binding to certain proteins on the sporozoite's cell surface, thereby marking the parasite for destruction by defensive cells (T lymphocytes) of human immune responses. In parts of Africa, some *Plas-*

modium populations have evolved genetic mechanisms that protect them from this immune response. Thus, malarial parasites and their hosts continually engage in an evolutionary arms race of genetic offense and defense, ploy and counterploy.

Traditional efforts to repress human malaria focused on mosquito control via insecticides and environmental measures (e.g., draining wetlands) and through the use of prophylactic and therapeutic drugs to prevent and treat *Plasmodium* infections. Although these efforts have been successful in many parts of the world, the disease has not been eradicated, and two troubling developments have surfaced in recent years: the evolution of insecticide resistance in some mosquito populations and the evolution of drug resistance in some populations of *Plasmodium*. This has led to calls for imaginative new approaches to malaria suppression.

Some researchers think that an answer lies in engineering transgenic mosquitoes for genetic resistance to *Plasmodium* infection. The basic idea is to introduce antiparasite genes into vector mosquito populations and thereby interrupt the protozoan's life cycle and thwart disease transmission to humans. The hope seems plausible for at least two reasons: "refractory" strains of some wild mosquito species are naturally incompetent to transmit the microbe, showing that genetic resistance is possible; and *Plasmodium* development inside the mosquito is highly complex, providing many potential target points for genetically disrupting the parasite's life cycle.

A major stumbling block, however, has been the lack of suitable experimental genetic transformation systems to introduce genes into mosquitoes. This technical hurdle recently was overcome. Using standard recombinant DNA techniques, alien transposable elements (see "Jumping Genes" in the appendix) were pasted into bacterial plasmids. The recombinant plasmids then were physically inserted into mosquito eggs, where the jumping genes naturally leaped into the mosquito chromosomes at random locations. Successful genetic transformation was documented because the jumping genes also had been engineered to express fluorescent markers (see "Reporter Genes" in appendix) that literally lit up the bodies of the transgenic mosquitoes.

A subsequent laboratory step was to engineer antiparasite genes into mosquitoes and coax the transgenes to express themselves in the insect gut. In one such experiment with *Anopheles gambiae* (the primary insect vector of human malaria in sub-Saharan Africa), GM mosquitoes transmitted 80% fewer *Plasmodium* sporozoites to mice than did non-GM control insects.

Next may come the truly big hurdles: successfully introducing these transgenic mosquitoes into nature and somehow getting them to replace their wild, infective counterparts. No one yet knows how this might be accomplished. It seems unlikely that sufficient numbers of transgenic mosquitoes could be mass reared and released to significantly impact natural populations, even if the survival and reproduction of the lab-raised strains were uncompromised. One theoretical possibility, however, is that the same jumping genes used to construct the transgenic insect strains might actively proliferate and jump around in nature too, thereby spreading antiparasite genes through natural mosquito populations like an infectious wave. But this is little more than sheer speculation at present.

In June 2002, an international conference held in the Netherlands highlighted the fact that whereas the molecular biology of GM approaches to malaria control has received great attention and research support, equally important questions on the ecological and social consequences of releasing vast hordes of transgenic mosquitoes into the wild have been almost totally ignored. This is a recurring theme in modern biology: Vast effort and money are expended on technical laboratory issues, while the ecological and environmental ramifications of the broader research program are neglected.

In any event, the simple act of developing a genetic transformation system for mosquitoes in the laboratory has been hailed as a technological breakthrough. It has some genetic engineers dreaming of a day when GM mosquitoes may block the spread not only of malaria, but of other mosquito-borne illnesses, such as dengue and yellow fever, too. But before such touted missions can be ranked higher on my boonmeter scale, major hurdles must be overcome, especially concerning the ecological issues of releasing GM mosquitoes into nature.

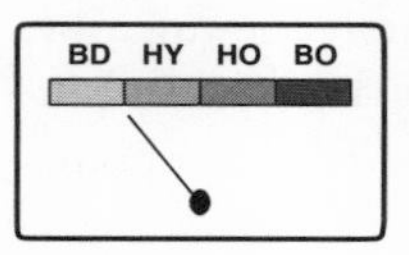

Fat, Sexy Salmon

In the early 1990s, biotechnologists astounded the world by genetically altering experimental coho salmon so that they grew many times faster than their wild relatives. A famous 1994 picture in the journal *Nature* showed five of these GM salmon lined up with five of their non-GM brethren, all at 14 months of age and reared under identical conditions. The contrast was startling. The transgenics were veritable behemoths, monsters weighing as much as 37-fold more than their puny nontransgenic counterparts. To some observers, the practical implications were clear. Biotechnology could revolutionize aquaculture industries by engineering transgenic fish for fast growth and high meat production.

The GM fish had been created by microinjecting salmon eggs with a molecular construct consisting of a fish growth-hormone gene attached to a promoter region that turned up production of the growth hormone to 40 times higher levels than in non-transgenic controls. The result was not only larger coho salmon, but also precocious sexual development as the GM fish began to mature at a younger age. Subsequent research revealed that the accelerated growth was associated with muscle hyperplasia (abnormal multiplication of cells) and several changes in patterns of gene expression.

Since then, the experimental introduction of growth hormone genes into fish has become something of a fad, having been accomplished also in chinook salmon, Atlantic salmon, rainbow trout, cutthroat trout, Arctic charr, mud loach, Japanese medaka, tilapia, flounder, and some 20 other species of fish worldwide. Part of the motivation simply has been to learn more about the physiological and metabolic action of growth hormones and to develop better laboratory techniques for genetic transformation in fishes. Another motivation, however, has been to improve profits from commercial fish farming.

In recent decades, with the dramatic decline of many native salmon stocks due to overfishing and habitat alterations by humans (such as dam construction), salmon farming has become big business in many parts of the world. Whereas native salmon complete their anadromous life cycle by migrating from the sea to spawn in freshwater streams, farmed salmon simply are raised to harvestable size in holding pens, large numbers of which are stationed along the coasts of Norway, Scotland, Japan, Canada, and other countries. Traditionally, these salmon have been non-transgenic, but the hope of many industry leaders is that growth-enhanced GM salmon will vastly improve the productivity and profits from these fish farms.

Even before the biotechnology era, ecologists harbored doubts about the merit of conventional fish-farming methods as applied to salmon. First, the net impact of salmon farming on total fish biomass in the sea is negative rather than positive, because penned salmon usually are fed manufactured fish meals and oils from other species, like menhaden and anchovies that already are being exploited at near maximum sustainable levels. Second, high feed inputs increase the accumulation of wastes that can promote eutrophication, sediment buildup, and other ecological problems near fish farms. Third, escaped farmed fish might introduce parasites or diseases into native salmon populations or unduly compete with their wild counterparts for limited resources. Salmon holding pens are notoriously leaky, so reared fish often escape into the wild. For example, an estimated 100,000 salmon recently bolted from fish pens in the Orkney Islands off the Scotland coast, and, in some spawning rivers of the northeastern United States, escaped farm salmon now outnumber wild salmon.

Another concern is that fish-farm escapees, which tend to be fairly homogeneous genetically, may interbreed with their wild relatives and thereby significantly contaminate native gene pools with "domesticated" DNA. Many native salmon populations already are listed as threatened or endangered. Will they be compromised even further if, by hybridizing with farmed animals, they lose their natural genetic instincts that tell them how to survive and reproduce in the wild? (Captive stocks of many salmonid species are known to differ genetically from wild populations, due in part to altered selective pressures associated with hatchery rearing.) Accordingly, some marine biologists urge that all farmed salmon should be sterilized. The techniques for doing so are relatively simple, although not always 100% effective. They involve shocking recently fertilized salmon eggs with pressure or heat (see "Daughterless Carp" below).

Does the prospect of rearing transgenic salmon in traditional fish farms raise any additional ecological risk factors? Some researchers say no, pointing out that even if some of the fat, growth-enhanced salmon occasionally escape from their pens, they would be inferior swimmers and otherwise unlikely to survive well in the wild. But recent theoretical models have cast considerable doubt on this cavalier attitude.

Individual survival is only a part of the equation fostering ecological success. Other key factors include female fecundity (the number of eggs produced), male fertility, age at first reproduction, success in attracting mates, and other reproductive capabilities. In salmon and many other fishes, these fitness components can be correlated with growth rate; large, fast-growing fish often mature at younger ages, are more fecund, and tend to be more attractive to other fish as spawning partners. Thus, in terms of net reproductive output, fat and sexy fish might outperform their svelte relatives, even if they themselves are somewhat debilitated physically.

By theoretically modeling the genetic fates of fat, but sexy, transgenic salmon versus skinnier native salmon under a variety of ecological and population-demographic conditions, researchers have explored this troubling possibility. According to their findings, a growth-hormone transgene introduced into a wild population by the escape of even a few farmed fish can, in principle, successfully invade a natural population and take over the native gene pool. One group of researchers called this the "Trojan gene" scenario. No one knows for sure that this would happen in real life, but it is a sobering and distinct possibility and provides one more reason to exercise great caution in any contemplated efforts to mass-farm growth-enhanced salmon.

Worldwide production of fish and shellfish amounts to more than 120 million tons sold annually, more than 75% of which is destined for human consumption. Nearly one-third of the latter comes from aquaculture, a proportion that is only likely to increase as more and more native fisheries collapse from overharvest. Clearly, much is at stake, both economically and environmentally, as fishing industries try to adjust to the new ecological and scientific realities of the twenty-first century. My own provisional conclusion for now, especially given the ecological uncertainties, is that large-scale farming of growth-enhanced GM salmon has been promoted with more hoopla than is warranted.

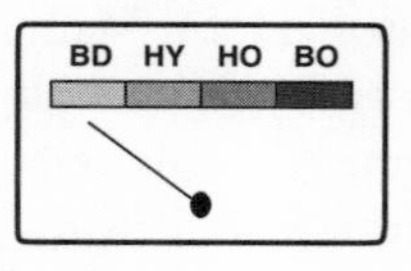

Antifreeze Proteins

Have you ever wondered how fish in the Antarctic Ocean survive the subzero temperatures that would freeze most other creatures? This question led scientists to make amazing discoveries about how some cold-blooded animals can withstand the most frigid of conditions.

Approximately 100 species of icefishes (Notothenioidea) thrive in southern polar regions courtesy of special antifreeze proteins (AFPs) circulating in their blood. About 15 million years ago, when the climate of a much milder Antarctic region began to chill, a gene that formerly encoded trypsinogen (a pancreatic protein) in ancestral icefishes began accumulating DNA mutations that eventually converted it to an antifreeze gene enabling the fish to withstand seas reaching –1.8°C. At this temperature, saltwater freezes, and most other marine fish have long since died. The glycoproteins specified by this gene do more to freeze-proof the icefish than glycol-based antifreezes do to protect automobile radiators. In addition to depressing the temperature at which fluids freeze, AFPs inhibit the growth of ice crystals and protect living tissues and cellular membranes from cold-induced damage. Indeed, the unique structural properties of AFPs make them up to 500 times more efficient than most other dissolved molecules in countering the effects of brutal cold.

Entrepreneurs salivate over commercial AFP prospects. They envision their use in such varied applications as providing a safe medium for cold-storing cells and organs for medicine, protecting healthy tissues during cryosurgery (wherein doctors use cold probes to freeze-kill solid tumors), preventing freezer-burn in meats and vegetables held at low temperatures for freshness, and improving ice cream so that it doesn't become fouled with ice crystals. In principle, such applications could be met by harvesting AFPs from the plasma of high-latitude fish, but this poses obvious logistic as well as ecological difficulties. It is estimated that ice-cream makers alone would need more AFP than could be supplied by 150,000 metric tons of polar fish!

To boost AFP availability, some scientists suggested inserting an AFP gene into microbes and then cajoling them to secrete the glycoprotein in fermentation vats. This idea recently passed one proof-of-concept test. A newly discovered AFP gene (from a high-latitude beetle, rather than a fish) was grafted into a laboratory strain of *Escherichia coli*, whereupon the bacteria dutifully produced the powerful anti-icing protein. The researchers suggested that such transgenic colonies could be scaled up to produce AFPs in much greater amounts than reasonably could be obtained from nature.

Other genetic engineering prospects for AFP are even bolder; they envision purposefully altering the freeze-tolerances of organisms in the outdoors. Scientists years ago succeeded in fitting tobacco, tomato, and potato plants with transgenes modeled after AFP genes from polar-region fish. After considerable recombinant DNA fiddling, some of the experimental plants were coaxed to express the AFP in ways that inhibited the formation of deadly ice crystals in the plants' tissues. Many plants of huge commercial importance (e.g., citrus trees) suffer damage from frost and cold, so it is easy to envision economic importance for GM crops that might be engineered to withstand chillier conditions. Of course, before such transgenic plants are released into the environment, people should think hard about ecological as well as economic ramifications.

Some plants also produce AFP-like proteins naturally. More than 30 species of flowering plants (including winter wheat, winter barley, and winter canola) are known to display antifreeze-protein activity, especially when acclimated to low temperatures. This sends mixed messages regarding GM prospects. On the one hand, species possessing natural antifreeze proteins surely are preadapted (metabolically and physiologically) for dealing with AFPs, so relatively little disruption of normal cellular operations might be expected if genetic engineers insert additional AFP genes or modify existing ones. On the other hand, AFP-carrying plants already have been under natural selection for freeze tolerance over the eons, so how much more genetic improvement in this direction might biotechnologists reasonably expect?

The boldest suggestion to date for AFP transgenes is also among the oldest: to genetically modify freeze-intolerant fish so that they too might thrive in colder climates. Primary experimental subjects have been salmon, which naturally succumb when water temperatures drop below –0.8°C. In 1982, scientists microinjected salmon eggs with more than a million copies of an AFP gene isolated from a polar species, the winter flounder. The transgene successfully integrated into salmon chromosomes, expressed properly, and was transmitted in normal Mendelian fashion through succeeding generations. These GM salmon also proved more tolerant of freezing conditions.

This finding carries potential economic significance because, if the GM fish are ever put into commercial production, salmon farming might be extended northward, well beyond, for example, its current limit in southern New Brunswick along North America's Atlantic Coast. As described in the preceding essay, some major ecological concerns attend commercial salmon farming, however, and these are likely to be magnified with the introduction of GM fish. Therefore, environmentalists and others are likely to retain a cold attitude toward attempts to engineer freeze tolerance into commercial fisheries operations.

Overall, genetic engineering with respect to AFPs is hard to evaluate, not least because of the diverse applications contemplated. Some of these, such as producing commercially viable quantities of antifreeze compounds from GM microbes, seem potentially beneficial and low risk. Others, such as imbuing GM crop species with novel AFPs, might provide huge agricultural benefits but also carry potentially negative ecological consequences. Still other GM projects with AFPs, such as engineering commercially harvestable fish species to withstand colder climates, seem to me to be pure boondoggle.

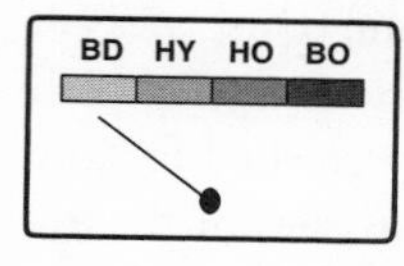

Thus, with particular regard to AFP projects that entail genetic engineering of species in natural environments (the topic of this chapter), I must give this topic a low score in the boonmeter.

Mutation-Detecting Fish

Mutations, the raw material of evolution, are a fact of life. However, only a small fraction of newly arisen (*de novo*) mutations actually increase an organism's prospects for survival and reproduction; most are harmful or at best merely fitness-neutral to their bearers. Thus, the mutational process would seem to be a necessary biological evil, permitting the continuance of the evolutionary game, yet leaving many ruined cells and lives in its wake.

Mutations arise spontaneously (often as errors during DNA replication, repair, or recombination) at low but consistent rates typically on the order of one in a million gametes per locus per generation. However, many physical and chemical agents, known as mutagens, enhance mutation rates. Among these are ultraviolet light, X-rays, and other high-energy radiations, plus various natural and synthetic chemicals such as nicotine, caffeine, arsenic, cortisone, and mustard gas. Most if not all mutation-causing factors are known to promote human cancers as well. Thus, there are valid reasons for concern when any mutagenic agent reaches high concentrations in the environment. Furthermore, more than 2000 new chemical compounds are introduced each year in an array of consumer products, and it is desirable to know whether any of these might elevate mutation rates appreciably.

Scientists have devised various ways to monitor mutation rates as biological indicators of the presence of mutagenic agents. A classic, widely used approach is the Ames test, in which bacterial colonies are exposed to varying concentrations of a suspected mutagen and specific marker genes then are monitored for *de novo* mutations that, under the experimental conditions employed, are revealed via predictable effects on bacterial growth. The Ames test permits rapid screening of vast numbers of bacterial cells, a logistic necessity given that mutations are relatively rare events even at the elevated rates that mutagenic agents might induce. A disadvantage of the Ames test is that bacteria probably experience the environment quite differently from multicellular organisms, including humans, the ultimate subjects of concern in most such toxicological studies.

Thus, for assessing mutagenic patterns more relevant to humans, a better model organism would be a mammal. Accordingly, scientists next developed a mutation-rate assay wherein specific bacterial genes first are introduced into laboratory rats using recombinant DNA methods. After these rats have been exposed to varying concentrations of a suspected mutagen, the transgenes are retrieved from the animals' tissues and reinserted into the bacteria from which they came. Then, as in the Ames test, these genes are mass-screened for any *de novo* mutations that arose in this case in various cells or tissues of the experimental rodents. The mutational tallying still is conducted in bacteria; however, many thousands of mutational targets that were present in the rat at the time of exposure can be screened rapidly by this approach.

A similar *in vivo* (within a live animal) test recently was developed for assessing mutation rates in the Japanese medaka (*Oryzias latipes*), a tiny freshwater fish. A phage or plasmid vector was used to insert specific mutation-detecting bacterial genes into medaka eggs such that all cells in the resulting GM fish contained the transgenes. The experimental fish then were exposed to a potential mutagen, and the transgenes later were rescued and again assayed for *de novo* mutations. The first known mutagen tested in this manner (ethylnitrosourea) increased mutation rates in medaka by as much as 10-fold above background levels. Furthermore, the mutation rates varied predictably according to magnitude and duration of chemical exposure, just as they did in conventional rat experiments, which validated this approach.

Why would anyone want to use transgenic fish as an *in vivo* model system for monitoring potential mutagens? There are several reasons. First, medaka and other tiny fish are easy to grow and manipulate. A fish costs only pennies a year to raise, compared to $0.20 or more a day per rat. Second, any mutational monitoring that uses scaly as opposed to fuzzy creatures seems to be more ethically acceptable to the general public. Third, fish live in water environments often contaminated by oil spills, heavy metals, herbicide and pesticide residues, and so on. By assessing mutation rates under such conditions, the hope is that transgenic fish might become useful biological monitors of aquatic or marine pollution. Such are the arguments advanced by geneticists now angling for increased funding to develop transgenic fish as environmental monitors of mutagenic agents.

However, important questions remain. Do the bacterial sequences actually targeted in such transgenic mutation assays accurately reflect the mutagenic response of the endogenous, non-transgenic DNA sequences that belong to the fish (or rats) themselves? Even if they do, are any observed increases in mutation rate actually biologically relevant to the health or survival of these vertebrate creatures, let alone to humans that likewise might be exposed to the same mutagen? More funding may indeed be required because definitive

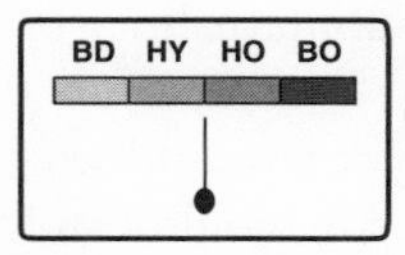

answers to these and related questions remain largely unknown. In short, it is simply too early to draw definitive conclusions about the prospects of GM fish for assessing potential mutagenic agents.

Sentinels of Aquatic Pollution

A more direct type of biomonitoring, also in its infancy, involves the use of live animals engineered to carry transgenes whose activation requires a specific chemical agent in the environment. In principle, if this particular chemical is present in an animal's surroundings and is taken up by the animal, a reporter gene linked to the activated transgene could then signal this fact to the researcher. Recently, this remarkable vision was experimentally implemented.

A good example involves zebrafish (*Danio rerio*), a small and easily reared fish that has the added advantage of possessing a translucent body. By microinjecting zebrafish eggs with plasmid-carried pieces of foreign DNA, scientists have created transgenic fish that house "pollution-inducible response elements." Each such element consists of a response gene that is activated naturally by a given physical or chemical agent, linked to a reporter gene that visibly luminesces or fluoresces when the adjacent response gene is turned on. This natural light is what the scientist then monitors and quantifies, either in cells or tissues recovered from the fish, or directly in the living fish (hence the relevance of a translucent body).

In effect, each inducible response element acts like a light switch that turns 'on' only in the presence of the relevant chemical agent (often a pollutant in the water). Indeed, the light switch operates more like a rheostat whenever, as often happens, the intensity of the luminescence also proves to be proportional to the chemical's concentration in the fishes' tissues. In such cases, the emitted light indicates not only the presence of the inducing factor, but also its approximate level.

At least six different inducible response elements have been developed in zebrafish to date, each apparently activated by a different environmental pollutant. For example, one such element is turned on by exposure to heavy metals such as cadmium, zinc, mercury, cobalt, or nickel. Some heavy metals are toxic, mutagenic, or carcinogenic, so it is important to know when they are present in a body of water or its sediments. Another genetic response element in transgenic zebrafish is induced by estrogenlike compounds that are common in pharmaceuticals and in environmental chemicals such as insecticides. At high concentrations, some such chemicals are suspected to disrupt normal cellular signaling and lead to reproductive and congenital abnormalities in animals.

The hope is that these transgenic zebrafish will prove to be powerful sentinels of water pollution, living biological sensors well situated to monitor and report on various chemicals of environmental concern. Basically, all the biologist needs to do is let the transgenic fish swim in the water for a while and see if their bodies light up. Each transgenic strain can be engineered with a different pollution-inducible response element, so an appropriate panel of such animals could serve as a school of "watchfish." In a preliminary field test of this experimental approach, transgenic zebrafish are being used as biomonitors for Ohio's Little Miami River, the east fork of which is downstream from a hazardous waste dump that some people fear might be leaching toxic wastes into the drinking water of Clermont County.

Biological monitoring for hazardous chemicals could have several advantages over conventional monitoring procedures. First, transgenic fish sentinels should be cheap and easy to monitor, requiring little more than a field luminometer (a device for quantifying light emissions) powered by a car or boat battery. Traditional laboratory methods of analyzing pollutants in samples of water, sediment, or animal tissue are far more expensive, labor intensive, and technically demanding. Second, some inducible response elements are activated inside a fish within a matter of minutes after exposure to the relevant chemical, so test results can be obtained quickly. Third, because the assays in effect are conducted by the fish themselves, results should be more relevant than traditional water or sediment assays to the true concentrations of pollutant chemicals within living tissues.

A fourth important point follows directly from the third. It has long been appreciated that fish and other aquatic creatures bioconcentrate many environmental contaminants, sometimes to levels that are tens of thousands of times higher than in surrounding waters. In such cases, what really matters to a living organism is the biologically effective concentration of each chemical in its tissues, and this is precisely what a pollution-inducible response element is designed to monitor.

The use of transgenic fish as pollution sentinels does have some potential difficulties. A major technical challenge has been to design fish that reliably transmit the proper expression of transgenes to their progeny. For reasons poorly understood, such expression thus far has been unpredictable. Another important technical challenge is to understand the potentially complex nexus of molecular events that can influence expression patterns of the sentinel genes. These may include, in addition to chemical concentration per se, such factors as the broader physiological state of the fish and interactive effects (positive or negative) among multiple agents in the gene-induction process. Thus, the bioassays will have to be standardized and carefully interpreted.

A final important point is that transgenic sentinel fish should seldom if ever be released into natural environments, where they might be disruptive to native biota and ecological processes. Fortunately, such biotic releases normally can be curtailed if the monitoring programs are conducted in closed aquaria containing the water or sediments to be tested.

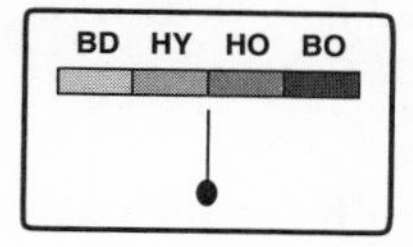

In conclusion, the use of GM fish as luminescent sentinels of aquatic pollution is still in its infancy. Thus, the boonmeter depicts my noncommittal stance on the prospects for this bizarre approach to establish and prosper.

Transgenic Environmental Biosensors

If transgenic fish can be engineered to serve as living sentries for aquatic mutagens and other pollutants, it should come as no surprise that genetic engineers plan to create and deploy terrestrial creatures in similar fashion. In the last century, miners often placed caged canaries in working tunnels to serve as early-warning alarms to impure air. Likewise, today's biotechnologists hope to deploy a wide variety of transgenic plants, animals, and microbes as sensitive biodetectors of cryptic and often hazardous foreign substances in the environment. This essay touches upon several examples under current research and development.

Sedentary plants, being closely tied to the land, should be favorable organisms for monitoring toxic soil chemicals, of which, unfortunately, there are many. Particularly since the Industrial Revolution, humans have burdened natural soils around the world with myriad pollutants ranging from complex organic poisons such as PCBs (polychlorinated biphenyls, used in pesticides), to simple inorganic metals that can be highly toxic in high doses, to radioactive elements. Now, biotechnologists are attempting to engineer plants to help detect and report such soil contaminants to humans.

Transgenic mustard plants (*Arabidopsis*) recently were designed to assay for heavy metals. Annually, humans dump more than 1.3 million tons of zinc, 950,000 tons of copper, 790,000 tons of lead, and 20,000 tons of highly toxic cadmium into the environment, for example. Much of this inorganic material contaminates soils near mines and various industrial sites. Some of the heavy metals, such as zinc and copper, are essential trace elements, necessary for life in miniscule quantities, and accordingly are added to many vitamin pills and processed foods. However, these and other metals become health risks at the much higher doses that characterize many polluted areas. When transferred to living tissues, heavy metals at elevated concentrations often exert their toxic effects by inducing damage to DNA molecules and cells and by promoting cancer, probably via their mutagenic actions.

The experimental mustard plants were engineered in such a way as to record mutations induced by heavy metals or other mutagens that might be present in soils. By inserting a transgene (for the enzyme β-glucuronidase) that permits a simple histochemical assay, new mutations could be detected and quantified in the plants' tissues. The researchers found that transgenic mustards exposed to increasing concentrations of heavy metals did indeed show significantly increased mutation frequencies, making the GM plants potentially useful as living biomonitors of soil pollution.

Transgenic mustard plants have also been used to scan for radioactive mutagens. On April 26, 1986, a meltdown at the Chernobyl nuclear reactor in the Soviet Union resulted in the release of a radioactive cloud that blanketed more than 600 km^2 of surrounding land with cesium-137, strontium-90, americium-241, and other radionuclides. Genetic studies of humans and indigenous animals near Chernobyl have indicated increased rates of germline mutations from the ionizing radiations, but plants would obviously be more suitable and convenient subjects for directed experimental analyses of the situation. Accordingly, researchers recently used GM mustard plants to assess the mutational potential latent in assorted Chernobyl soils.

Such living sensors of mutagenic agents in the environment have the advantage of revealing bioavailable levels of pollution relevant to plant or animal health. However, other transgenic detection systems may be far simpler and more straightforward. In Aberdeen, Scotland, a startup company, Remedios, has developed GM bacteria that glow yellow in clean soils, but not when the soils are contaminated by particular chemicals. By using a battery of pollution-sensitive microbes carrying luminescent transgenes derived from fireflies or marine bacteria, the company can screen for volatile organic solvents like benzene, toluene, and xylene, for toxic nonvolatile organic compounds like PCBs, and for toxic inorganic substances like heavy metals.

Another potential use of transgenic microbes is in landmine detection. Some bacteria can sense TNT (trinitrotoluene) or other explosive chemicals, and scientists would like to capitalize on this inherent ability to help find thousands of hidden landmines that cause horrific carnage in former war-torn regions of the world. The idea is to fuse a fluorescence transgene to a bacterial TNT-sensing gene and then spray (e.g., by crop-duster planes) the GM microbes over suspected minefields. Landmines leak small amounts of TNT over time. Thus, in any transgenic bacteria that fall near a mine, both the TNT-sensing gene and its adjacent reporter transgene become activated. By then shining an ultraviolet light over the affected landscape at night, the precise location of each landmine is revealed as a fluorescent spot in the soil. In experiments with this approach, transgenic bacteria located all five buried sources of TNT in a 300-meter test plot in South Carolina.

Sometimes, fluorescent reporter genes are engineered into creatures for the purpose of identifying the transgenic organisms themselves, rather than abiotic substances in their environment. One example involves cotton bollworms. As described earlier (see chapter 4), one traditional method of controlling this insect pest is to release large numbers of laboratory-sterilized individuals into nature to compete with fertile native moths for mates. Once released, however, the sterile individuals normally cannot be distinguished from the wild animals by simple visual inspection. So, using recombinant DNA techniques, researchers added a luminescent transgene to some lab-reared batches of sterile bollworms. These animals fluoresce under UV light, a property that could assist in future efforts to monitor the abundance and distribution of released specimens.

The fluorescent GM organisms and their applications presented thus far may seem bizarre, but one final example is genuinely otherworldly. Scientists recently announced plans to use tiny transgenic plants to monitor environmental conditions on Mars. The National Aeronautics and Space Administration has scheduled a 2007 mission to explore Mars, and onboard the spaceship may well be mustard seeds engineered with reporter genes designed to incandesce if particular substances are present. After their 286-million-mile journey, the seeds would be planted in Martian dirt (and presumably watered, fertilized, and tended) by a gardener robot, and a camera would record the outcomes. Any germinated plants exposed to heavy metals would turn fluorescent green, for example, and those exposed to peroxides would turn blue. Such experiments might add a bit more color to Mars, as well as cast new light upon its surface conditions.

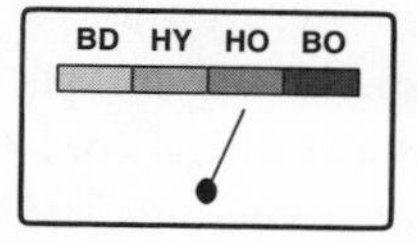

Phytoremediation of Mercury Poisons

Biotechnologists are keen to do more than biomonitor for pollutants; they also want to clean up the environment. There is plenty to be done. The Environmental Protection Agency estimates that at least 30,000 polluted sites in the United States are in need of remediation from toxic wastes. In the United Kingdom, 100,000 contaminated sites are known or suspected, affecting about 1% of its total land area. These and untold numbers of other sites around the world are polluted with witches' brews of routine organic and inorganic compounds, an unhappy legacy of decades of industry, mining, unwise agricultural practices, and other human activities.

Various heavy metals that have been released into soils or waterways are particularly troublesome for several reasons. First, when transferred at high doses to living tissue, they can be teratogenic (i.e., induce developmental malformations) in fetuses, as well as toxic or carcinogenic (cancer-causing) at any life stage. Second, unlike many organic compounds, heavy metals cannot be broken down into nontoxic subunits (although some can be modified into less harmful forms). Finally, heavy metal residues are widespread in the environment, originating from such diverse sources as leaded paints and gasolines, scrap metals, expended ammunition, power plant wastes, and industrial chemicals (leaked, discarded, or abandoned).

The dangers of heavy metals as well as the opportunities they present for bioremediation are well illustrated by mercury (Hg). This toxic element enters the environment either in volatile liquid form [metallic Hg(0)] typically from industrial accidents, or in nonvolatile particulate form [ionic Hg(II)] in dregs from trash, burnt coal, or commercial chemicals (as well as from natural volcanic activity). Any form of mercury can be toxic, but the most serious problems usually come from methylmercury (MeHg), a compound naturally produced by anaerobic bacteria when their habitats are polluted by metallic or ionic mercury. Methylmercury is a potent toxin because, as it moves up food webs into flesh-eating animals (including humans), it becomes bioconcentrated by several orders of magnitude.

The world was alerted to the danger of methylmercury by a disaster on Japan's Kyushu Island in the 1950s. For years, a chemical factory had dumped wastewater laden with mercury into Minamata Bay, resulting in the death of many shellfish, fish, and birds. Nobody paid much heed, however, until cats began to convulse and die in droves, and local villagers experienced an epidemic of sensory disturbances and brain lesions. Even as the chemical company (Chisso Ltd.) denied responsibility and the Japanese government looked the other way, methylmercury from ingested seafood was confirmed as the cause of "Minamata disease." In the ensuing years, more than 12,000 people living near the bay experienced symptoms of this disorder, 2252 were clinically diagnosed with the illness, and more than 1000 cases were fatal. Later, the government spent 50 billion yen over a decade in attempts to clean up this environmental mess, but MeHg still pollutes the bay, and fishing bans remain in effect.

How might biotechnology help clean up mercury spills? Researchers at the University of Georgia think they have an answer in transgenic plants engineered to detoxify mercury's more poisonous forms. Experimental mustard plants were modified to carry two genes of bacterial origin: *merA*, which converts Hg(II) to Hg(0), and *merB*, which degrades MeHg to methane and Hg(II). Hg(0) is a relatively benign form of mercury [100-fold less toxic than Hg(II)] that the GM plants either sequester in their tissues or volatilize into the atmosphere as they transpire water vapor through surface pores. Due to their newly acquired metal-detoxification systems, these transgenic plants can withstand levels of soil mercury about 10–50 times greater than those that kill non-GM controls.

Of course, the intent is not to produce mercury-resistant plants per se, but rather to enlist GM plants to cleanse contaminated soils or waters. The idea is that appropriate transgenic plants will draw up MeHg or Hg(II) through their roots, process these poisons to Hg(0), and release the elemental mercury harmlessly into the air. Whether this approach will work on large outdoor scales required for environmental restoration remains to be demonstrated, but the promise appears great.

Mercury is just one of many environmental pollutants projected for ecological clean up using GM plants. Cadmium, cesium, chromium, lead, selenium, strontium, tritium, and uranium are other poisonous waste-site elements that engineered plants someday might help decontaminate. So, too, is arsenic, which contaminates drinking water in Bangladesh and parts of India and causes health problems for an estimated 110 million people. Recently, researchers moved arsenic-processing genes (that absorb and bind arsenic from soils) from bacteria into experimental plants. The hope is that such GM plants can be mass planted in arsenic-contaminated areas to help sequester this environmental poison.

Unicellular GM plants are also being recruited to such tasks. Algae (like other species) require trace quantities of various minerals, and scientists are taking advantage of this natural proclivity. By tweaking and shuffling responsible genes and metabolic pathways, researchers are making algae better suited as hyperacculumators of undesirable elements at polluted sites. For example, GM strains of *Chlamydomonas* algae are being engineered to clean up trace elements such as mercury, cadmium, and zinc from contaminated sediments in Lake Erie. One promising approach is to attach metallothioneins (proteins that bind heavy metals) to the outside of the algal cells so that they scavenge these toxins more efficiently.

A related prospect is to engineer plants to bioconcentrate valuable metals for mining purposes. When minerals such as copper or zinc are taken up from soils and sequestered in plant tissues, they might then be harvested at less cost and less environmental damage than traditional extraction methods such as strip mining.

Phytoremediation (environmental restoration by plants) could provide an ecologically friendly alternative to current pollution cleanup methods. Once in operation, the process would be sun-powered and potentially low in cost (relative to conventional remediation techniques such as soil excavation and removal). Phytoremediation also has some limits, however: It is slower than soil excavation, it can clean soil only to the depth of the GM plants' roots, and there is still the problem of how to harvest and dispose of the metal toxins sequestered in the GM plants' tissues. Other concerns are that the metal-sequestering transgenes might escape (via hybridization) into wild or cultivated relatives of the engineered plants, and that the toxin-laden plants might be eaten by wildlife, either harming the herbivores directly or by harming other species up the food chain.

In a broader sense, phytoremediation, promoted by humans, is nothing new. For centuries, farmers have rotated crops with leguminous plants to fertilize nitrogen-depleted soils, foresters have seeded barren landscapes to control erosion, and gardeners have planted aquatic vegetation to oxygenate and filter polluted waters. From this perspective, genetic engineering is just a new-age way to enlist plants to modify environments for human purposes.

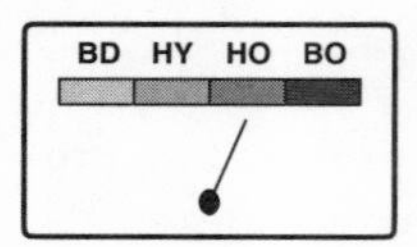

Phytoremediation of Organic Pollutants

Many plants come naturally well suited as environmental monitors and cleansers. Their root surfaces often are extensive and chemically sensitive, having evolved specifically to adsorb (accumulate on the outside) or absorb (take inside) nutrients from soils and waters. Most plants have innate systems for translocating, sequestering, and sometimes deporting (e.g., via transpiration or leaf fall) toxic foreign compounds. Plants collectively have huge biomass, making them potentially useful as repositories and/or detoxifiers for vast amounts of pollutants. Plants in effect are "self-powering" systems, equipped with solar panels (leaves) that collect energy from natural sunlight. Finally, being sedentary and dependent on local environmental conditions, many plants have evolved genetic capabilities to detoxify or otherwise ameliorate the poisonous effects of numerous site-specific contaminants.

By genetically tweaking plants for improved adsorption, uptake, transport, storage, degradation, or volatilization of environmental toxins, biotechnologists are now attempting to upgrade plants' natural proclivities for bioremediation. This experimental field is still in its infancy, but already it has spawned a lexicon of fancy terms whose general meanings nonetheless are rather evident: phytoextraction, phytoaccumulation, phytostabilization, phytodegradation, phytovolatilization, and rhizofiltration (which is identical to phytoextraction except that the plants' roots grow in water rather than in soils, to remove pollutants directly from aqueous environments).

The previous essay described experimental attempts to remove toxic inorganic substances from contaminated sites. Analogous approaches are being explored to phytoremediate organic pollutants, many of which fall into four broad categories: petroleum-based products, specific nitroaromatic compounds such as the explosive TNT, chlorinated poisons such as PCBs, and an assortment of other toxic residues often traced to chemical pesticides. Many organic compounds can be phytodetoxified by one additional route: degradation into simpler and less-toxic substances via a plant's basic cellular metabolism. Below I present three areas of ongoing or contemplated research that illustrate how biotechnologists aim to enhance these metabolic pathways through transgenic technologies.

Trichloroethylene (TCE), a toxic and carcinogenic industrial solvent, is a widely distributed pollutant of groundwaters and soils. Native plants grown at polluted sites long have been known to extract and transpire (emit as a vapor) TCE and to promote the growth of root-associated bacteria that degrade the compound. A more recent discovery is that poplar trees, *Populus*, also produce enzymes that break apart TCE into simple chlorine compounds and harmless carbon dioxide. As more is learned about the mechanistic details and the taxonomic distribution of this degradative pathway, it may become possible for genetic engineers to enhance the trees' own detoxification processes or to transfer them into other plant species perhaps more suitable for remediation efforts (e.g., by virtue of being faster growing).

A second research example is somewhat farther along. As discussed in "Transgenic Environmental Biosensors," some bacteria can detect explosive compounds such as TNT, and genetic engineers have built upon this native ability by fusing reporter transgenes to the bacterial sensor genes such that the microbes fluoresce when grown in contact with landmines. Several plant species likewise can degrade TNT and related explosive compounds such as nitroglycerin, albeit at modest efficiencies. Recently, a nitroreductase gene was isolated from bacterial colonies and inserted into tobacco plants. These transgenic seedlings tolerate TNT and nitroglycerin far better than do non-transgenic controls and apparently break down nitroglycerin at about double the normal pace. This gives a substantial boost to prospects that plantations of transgenic plants may someday help decontaminate the thousands of acres of polluted land and rivers near sites that produce, store, and dispose of munitions.

A third example involves PCBs, among the world's worst pollutants due to their acute biotoxicity, wide distribution, and long half-life in the environment. Some plants degrade PCBs with modest effectiveness, as do various bacterial species that collectively have dozens of genes for this task. Using recombinant DNA techniques, biotechnologists hope to wed various microbial and plant genes, insert the genetic constructs back into plants, and sow those GM strains that offer the best prospects for phytoremediation of PCB-contaminated lands.

The preliminary state of such examples illustrates that much more research is needed into the basic genetics and metabolics of detoxification systems used by diverse plants and microbes. Other hurdles also must be cleared before GM plants can live up to their tremendous potential in decontaminating our planet of pollutants. In particular, issues of health, safety, and containment must be addressed because, almost by definition, any transgenic plants engineered for ecological restoration will have to be released into the environment to do their duty.

In nearly every bioremediation project involving GM plants, questions of the following sort will arise. Can any toxic compounds accumulated in plants' tissues be harvested and disposed of safely (if they are not already detoxified by the plants themselves)? Might the transgenes escape from the site of release, or transfer to other species, and with what consequences? Might some of the transgenic plants themselves disrupt the ecological communities they enter? There may be risks, but the health and well-being of humans and natural ecosystems already have been compromised by the toxins and pollutants that we have spewed into the environment. Thus, another ultimately compelling question for society will arise as well: Do we have any rational choice but to act?

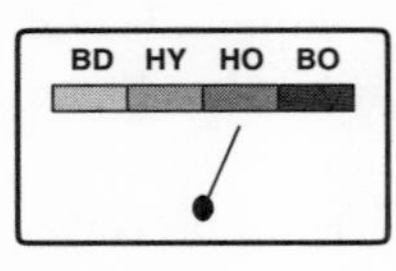

Salt and Drought Plants

Another form of environmental remediation is farmland restoration. Since the 1940s, unwise farming practices around the world have led to a substantial reduction in the plant-growing capacity of an acreage roughly equal to China plus India and the almost hopeless ruination of an area the size of Hungary. To return such land to productivity could go a long way toward alleviating world hunger without jeopardizing wild lands not yet cultivated.

Much of this farmland degradation is due to salt accumulation in soils. When crops are irrigated over the years with mineral-laden water, continued evaporation can lead to the deposition of calcium and magnesium salts that are toxic to most plants. Some estimates suggest that 20% of all cultivated land in the world, and nearly 50% of irrigated farmlands, have been harmed by such salinity increases.

Soil erosion is another major problem. Every year, globally, plowed farmlands lose an estimated 25 billion metric tons of topsoil to wind and water. Some of the most dramatic episodes are associated with periodic drought. For example, throughout the 1930s, a large mid-section of the U.S. southern interior, formerly a wheat-growing region, became a desolate dust bowl from a combination of dry conditions, overgrazing, and overplowing. Winds whipped across the desiccated fields, raising billowing dust clouds that carried away the precious topsoil, blackened the skies for days, and resulted in a mass exodus of thousands of farm families from the barren land.

Through plant genetic engineering, some scientists think that they can help societies avoid such disasters, as well as reclaim farmland that already is degraded. Research progress to date is limited, but nonetheless worthy of mention both for its considerable promise and potential pitfalls.

One suggested approach is to engineer salt-tolerant plants, and at least one such example already exists. In 2001, reports appeared concerning the construction of GM tomato plants that can grow in highly saline soils. Scientists engineered this feat by implanting regular tomato plants with a transgene encoding a membrane transport protein hooked up to a promoter sequence from a virus. In the GM plant, this genetic construct proved to pump salt ions from the soil, via roots and tissues, into cellular storage sacks known as vacuoles. The GM tomato plants were able to grow in soils nearly one-third as briny as seawater and 50 times saltier than their non-GM brethren can tolerate. Salt accumulated in the leaves, so the fruit reportedly tasted normal. If such GM plants can be commercialized, they promise not only to make salt-polluted acreage more productive, but also to rehabilitate such farmland to its original, functional low-salt condition.

A related approach showing promise is to engineer crop plants for drought tolerance, both to mitigate chronic water shortages (probably the biggest single threat to world food production) and to ward off singular disasters like the Dust Bowl. Several genes recently have been identified that naturally help plants cope with arid conditions, as well as with other stresses such as cold and salinity. Drought damage, salinity injury (which occurs when roots can't extract enough freshwater from salt-laden soils), and harm from frost (which occurs when water seeped from cells forms ice crystals in intercellular spaces) are all, in effect, symptoms of plant dehydration. By tinkering with various promoter sequences and genes and inserting the genetic constructs into cotton, tobacco, tomato, potato, canola, and others, researchers are trying to imbue several major crop species with the ability to tolerate desiccation. The Rockefeller Foundation in New York recently committed up to $50 million to this effort by supporting a decade-long global effort to improve drought resistance in GM maize and rice, for example.

If successful, such genetic modifications would have obvious benefits, but also some potential environmental dangers that sometimes get neglected in all the ballyhoo. Crops engineered for greater tolerance to salt, cold, or drought likely would be planted not only on reclaimed farmland, but also in areas such as ocean margins and deserts previously deemed uncultivable. In such cases, what constitutes "environmental remediation" may be in the eye of the beholder. Extensive wheatlands covering the Mojave Desert, for example, would not be everyone's idea of an environmental improvement.

The global quantity of arable real estate (land devoted to crops) is now 3.7 billion acres, or about 0.6 acres per person. Using current projections on population growth and cultivable land, by the year 2050 this likely will decline to about 0.3 acres per capita, an amount deemed insufficient by the

United Nation's Food and Agriculture Organization to adequately feed many of the world's poor, given current per-acre crop yields in numerous developing countries. To address this situation, three supposed solutions routinely crop up: boost farm production per unit acreage, increase arable land, or both. The jury is still out on the extent to which GM plants may improve crop yields on existing cultivable lands (chapter 4). The jury has barely convened as to whether plants engineered for salt and drought tolerance will significantly increase arable land or as to what broader environmental ramifications might be entailed in such activities.

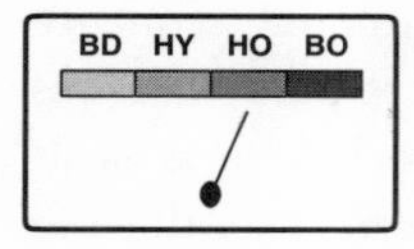

Bioremediating Bacteria

Even more so than plants, many bacterial species are well suited for bioremediation efforts. First, they display a variety of genes and metabolic pathways enabling them to utilize a wide spectrum of natural inorganic and organic compounds, many of which are structurally related to man-made chemicals. Second, they can divide and multiply rapidly, thus offering a large biomass for detoxifying environmental pollutants. Third, these microbes reside in abundance in nearly every kind of habitat, including soils and waters that are traditional dumping sites for xenobiotic (non-native) pollutants from human activities. Finally, bacteria have small and often well-characterized genomes that are favorable targets for genetic engineering.

Over the eons, bacteria have evolved genetic systems for extracting nitrogen, carbon, and energy from native compounds in their environs, but seldom do these metabolic pathways work with great efficiency on synthetic poisons or other recalcitrant industrial chemicals that societies now want to cleanse from polluted soils and waterways. High on this contaminant list are chlorinated compounds such as PCBs, commercial hydrocarbons like benzene and toluene, nitroaromatic compounds such as TNT, and heavy metals and radioactive elements. Having seldom or never been exposed to high levels of many such substances, most bacteria tend not to deal effectively with them.

However, by genetically modifying chemical-conversion functions already present in various bacterial species, biotechnologists are aiming to improve microbial performance with respect to pollution cleanup. For example, bacteria, as well as other organisms, collectively carry genes for hundreds of different cytochrome P450 enzymes that perform diverse chemical functions such as cleaving carbon–chlorine and ether bonds. Like a customized baseball mitt, each P450 molecule has an indented pocket that grasps a specific class of chemicals for processing. Via site-directed mutagenesis (see "Knockouts and Resuscitations" in the appendix), gene shuffling, and other genetic engineering methods, the shapes or sizes of these pockets can be modified in specific ways that alter the enzyme's substrate specificities and catalytic efficiencies. For example, in one experiment using the bacterium *Pseudomonas putida*, geneticists modified a P450 protein to convert polychlorinated benzenes to less noxious compounds. Another research project increased the microbe's capacity to degrade tetrachloroethane, an environmental pollutant used by industries as a solvent and an intermediate in the production of other chemicals.

At least some substances targeted for bioremediation, such as heavy metals and certain inorganic compounds, are not strictly xenobiotic; they are present in nature as well, albeit seldom in the amounts often found at waste sites. Having been exposed to low or modest concentrations of such toxins over the ages, many microbes have evolved decontamination systems that may convert the substance to less dangerous forms (biotransformation), isolate the substance at the cell surface (bioprecipitation), or cloister the substance at safe locations inside the cell (biosorption). Scientists are now tinkering with the genes and pathways engendering such capabilities. For example, they have engineered bacteria to hyperexpress metal-recognition proteins on cell surfaces such that the GM microbes now adsorb five times more zinc, lead, and nickel ions than their non-GM relatives. In another experiment, scientists engineered a bacterial strain with higher expression levels of metallothioneins, a large family of microbial proteins that specifically sequester heavy metals.

Mixed-waste sites will offer some of the stiffest cleanup challenges for GM bacteria. These sites typically contain complex cocktails of organic and inorganic poisons, often in conjunction with radioactivity. In the United States, for example, more than 1000 chemical waste sites are radioactively "hot," often the result of unwise discharges of nuclear-weapons wastes especially from the mid-1940s to the mid-1980s. The Los Alamos National Laboratory (which engineered the atomic bomb) recently embarked on an ambitious program to custom-engineer bacterial proteins for the express purpose of cleansing the environment of such pollutants.

Similar efforts are underway at many universities and industrial labs, with some initial successes. For example, one consortium of researchers put together a battery of bacterial genes for the potential treatment of toxic ionic mercury [Hg(II)] and poisonous toluene in highly radioactive waste dumps. They started with a natural soil bacterium, *Deinococcus radiodurans*, that can withstand doses of radioactivity that debilitate or kill most other organisms. Into this remarkable microbe, which is among the most radiation-tolerant creatures on earth, scientists genetically inserted a mercury-processing gene (*merA*) as well as a toluene-metabolizing genetic construct, from other bacteria. The GM *Deinococcus* proved to be able to convert Hg(II) to less toxic elemental mercury and simultaneously degrade toluene, even in high-radiation settings.

Research on genetically modified microbes (GMMs) for ecological restoration is still mostly at the developmental stage, and precious few of the GMMs have been field-appraised to date. To fully exploit GMMs for all they are worth in bioremediation efforts will be technically challenging and expensive. However, if GMMs can safely be engineered and deployed to clean up the toxic messes that we have created, their total return on investment will be a bargain compared to the estimated $250–400 billion price tag on pollution cleanup (in the United States alone) by physical soil removal and other traditional means. Thus, hopes are high for this technology.

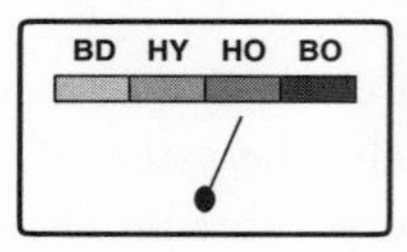

Cries over Spilled Oil

Near midnight on March 24, 1989, a huge oil tanker struck Bligh Reef in Prince William Sound, Alaska, disgorging 11 million gallons of crude oil into pristine waters and intertidal zones. The *Exxon Valdez* spill sparked a public outcry as sea otters, birds, and other marine life died gruesome deaths before TV cameras and a beautiful coastline was desecrated by black toxins. Emergency efforts to mitigate the disaster by conventional means—mechanical booms and skimmers, chemicals sprayed to disperse the pollution, and a fire purposefully lit to burn the crude oil—were thwarted by bad weather conditions and the sheer volume of the spill.

The *Exxon Valdez* was the largest maritime oil spill in U.S. history, and it prompted Congress to pass the Oil Pollution Act of 1990 that tightened oil-tanker industry regulations. But in terms of petroleum volumes, the Alaskan catastrophe at that time ranked only 28th on the all-time global list. The largest discharge, in 1979, was a 78-million-gallon spill from the *Atlantic Express* in the West Indies. Other infamous headliners were the 1978 *Amoco Cadiz* disaster along the French coast (68 million gallons), and the 1967 *Torrey Canyon* calamity in the English Channel (38 million gallons).

Other petroleum disasters have started on land. In March 1993, 400,000 gallons of oil spewed from a ruptured pipeline in Fairfax County, Virginia. In January 1988, a 4-million-gallon storage tank split apart in Floreffe, Pennsylvania, sending out a surge of oil that contaminated the Monongahela and then the Ohio rivers, killing wildlife, and for a time debilitating the economic, as well as the environmental, health of the region. As devastating as such local incidents can be, the total volume of oil that all such disasters belch into the environment pales in comparison to the chronic global release of 500-million gallons, annually, from nonaccidental sources such as routine road runoff and vehicle maintenance. Sooner or later, most leaked oil ends up in oceans.

Native soil and water bacteria are able to break down most petroleum products, eventually to carbon dioxide, water, and microbial biomass, but the process is too slow to have significant quick impacts on serious oil spills. To enhance such microbial bioremediation, two traditional approaches are used: fertilization and seeding. The first method, also called nutrient enrichment or biostimulation, involves adding phosphorus, nitrogen, or other substances to foster the growth of microbes capable of oil degradation. The second method, also called bioaugmentation, involves adding nonindigenous bacteria to boost the oil-degrading rate of the resident microbial population at polluted sites. Although both are widely used for U.S. oil spills, neither method is hugely effective in shoreline cleanup, especially on wave-tossed shorelines such as the coast where the *Exxon Valdez* foundered.

Some biotechnologists think they can genetically engineer microbes for improved oil-cleaning performance in nature. Indeed, the first bioremediation field trial for any genetically modified microbe involved a bacterium modified to better degrade naphthalene, a petroleum product. The process began with natural strains of *Pseudomonas fluorescens* isolated from soils at a gas-manufacturing plant. In the laboratory, plasmids were constructed and used to transform the parent strain so that it contained genes for naphthalene degradation fused to a bioluminescent reporter gene. The procedure was done in such a way that the GMM emitted bioluminescence in direct proportion to its rate of naphthalene degradation, thereby providing a built-in visible monitor of its performance. When released into contaminated subsurface soils, the GMMs remained viable and gradually did their duty, breaking down naphthalene pollutants.

An additional problem, salinity, attends efforts to engineer bacteria for the degradation of oil and other hydrocarbons in the marine realm. Most known oil-eating bacteria have been isolated from terrestrial environments and thus do poorly or die when exposed to the salty conditions of estuaries and shorelines. So, another group of geneticists recently engineered strains of *Pseudomonas* to contain osmoregulation genes that permit the bacteria to live in hypersaline environments. The stains already possessed a genetic capacity to attack various chemical fractions of crude oil, so the new GMMs, it is hoped, someday may help clean up oil disasters in and near oceans.

Regardless of any foreseeable success for remediation biotechnology, full restoration from major oil spills will remain a long-term process, and ecological harm will be done in the interim. Thus, even if GMMs someday do prove to be genuinely helpful, it will always be wise to focus primary efforts on preventing such spills in the first place.

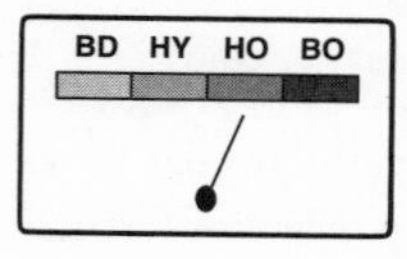

Rabbit Contraception

In the mid-1800s, English rabbits, *Oryctolagus caniculus*, were imported to Australia as a convenient source of meat, but some of these mammals escaped from their hutches, multiplied like rabbits, and their populations ran wild. More than 600 million rabbits soon overpopulated the continent, devastating croplands, pastures, and forests with their voracious appetites and penchant for digging burrows.

The problem became so great that Australians fought back with every available tool. Early pictures show thousands upon thousands of rabbit pelts, often set out to dry in long racks, from animals that had been shot, trapped, or poisoned. Another approach was to erect rabbit-proof fences. Some of these surrounded single farms, but others stretched for hundreds of miles in attempts to thwart expansions of the exploding rabbit populations from one Australian region to another. None of these pest-control methods worked particularly well.

The desperate situation called for drastic action, and in 1950 a bold approach was initiated. The myxoma pox virus, a natural disease agent of rabbits, was intentionally introduced into Australia from Brazil, the purpose being to create a myxomatosis epidemic. This oft-fatal sickness of rabbits and hares (family Leporidae) is spread primarily by fleas and mosquitoes, and this illness can be devastating to naive leporid populations not previously exposed to the virus.

The experiment worked, initially: Rabbit populations plummeted as the pox virus spread across Australia. However, the most virulent strains of the virus died with the rabbits they infected, leaving only the temperate forms in the rabbits that survived. Concurrently, rabbit populations naturally were selected for increased viral resistance. What emerged was an evolutionary standoff in which the surviving rabbits and the pox virus came to coexist in quasi-balance. Rabbit numbers in Australia are far lower than in the pre-pox era, but the feisty mammals still thrive and make nuisances of themselves.

For pest species like rabbits that proliferate rapidly and occupy large geographic areas, control strategies that focus on mortality factors alone are usually less than fully effective. In Australia, this has led to calls for alternative approaches that might impact rabbit fertility rather than survival. The result has been considerable scientific experimentation on rabbit contraception, and this is where genetic engineering has entered the picture.

In female mammals, eggs and early embryos are surrounded by a glycoprotein layer known as the zona pellucida (ZP). This extracellular coat plays a key role in reproduction by acting as a gatekeeper for the passage of fertilizing sperm and by influencing embryo implantation in the womb. It has long been known that if the immune system of a female develops antibodies against one or another glycoprotein in the ZP, her fertility can be diminished or blocked. Called "immunocontraception," this phenomenon has been exploited to develop injectible antifertility vaccines for more than 90 mammalian species. These dart-delivered vaccines have been effective in controlling the proliferation of large mammals, such as elephants, that may overpopulate zoos or natural landscapes, such as some parks in southern Africa. However, for logistical reasons, injected vaccines are of little use against abundant small mammals in the wild. Recently, however, Australian scientists conceived of a new way to deliver immunocontraceptives to wild rabbits: via genetically engineered viruses.

In one such procedure, a ZP-coding gene was inserted into a strain of myxoma virus. When the virus was allowed to infect experimental rabbits, the ZP glycoprotein not only was expressed, but it also elicited an immune reaction that significantly reduced the rabbits' fertility. Other viruses that specifically infect rodents have been engineered to confer partial sterility in house mice. Analogous kinds of genetic engineering can be envisioned that in effect might cause partial or complete male sterility, for example by eliciting female antigens against sperm or by altering sperm-expressed proteins that interact with the ZP or egg during fertilization.

Conceptually, genetic engineering for immunocontraception provides an interesting departure from conventional experimentation with most other GM traits in nonpest species. Normally, transgenes are engineered with the goal of improving an organism's performance (e.g., in growth rate or in resistance to diseases). The hope is that progeny of the initial GM generation will inherit the transgene and also benefit from it. In contrast, individuals engineered for sterility cannot transmit transgenes "vertically" because, by definition, they have no offspring. Thus the need arises for a delivery system that can spread sterility genes "horizontally" in the pest population. In the Australian case, the recombinant myxoma virus provides the infectious delivery vehicle intended to drive the rabbit population down.

In principle, genetically engineered sterility has some key advantages over traditional methods of mammalian pest control. To animal rights advocates, immunocontraception promises a humane alternative to overt killing by guns or poisons. To farmers and ranchers, the procedure might afford control over devastating feral pests that otherwise have proved impossible to eradicate. To ecologists, immunocontraception might offer an environmentally friendly alternative to chemical agents for eliminating pests.

However, some salient issues must be resolved before any mammalian contraceptive transgenes are released widely into the environment (by viruses or via other delivery systems such as transgenic plants). First and foremost, there must be strong assurances that a live vector does not reach and sterilize nontarget species. The myxoma pox virus is thought to be specific to Leporidae, but the possibility of dangerous transgene movement to unintended hosts cannot be neglected. A related concern is that the live myxoma virus might escape from Australia and harm rabbits elsewhere. Finally, the entire approach could fail if rabbit populations in Australia evolve genetic resistance to these contraceptive tactics.

To circumvent the first two problems mentioned above, alternative means to lower fertility have been contemplated. For example, contraceptive compounds could be delivered to rabbits via darts, baited foods, or by attenuated (dead or debilitated) virus. But at best, such halfway contraception would yield halfway success, because the sterility would be noninfectious and would impact merely the relatively few animals that could be reached by such methods of direct application.

Feral rabbits, rats, cats, goats, and other small mammal species are often devastating to the native lands and biotas they invade. Furthermore, no easy way exists to eliminate such pests once they have been unleashed into a natural ecosystem. Thus, in the future, far more effort should go toward blocking introductions of alien species, rather than dealing with the consequences after the fact. With respect to eliminating exotic pests, a few ounces of prevention can outweigh many pounds of attempted cure.

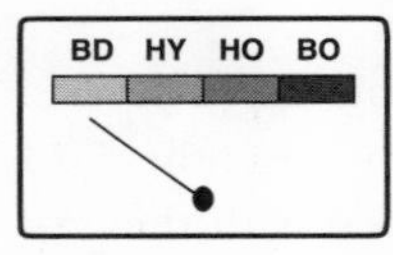

On balance, due to the exceptional dangers entailed, relative to the potential gain of lowering rabbit numbers in Australia, I gauge this project as a current boondoggle.

Daughterless Carp

Terrestrial creatures like rabbits and mice aren't the only invaders plaguing Australia (see previous essay). The continent's aquatic habitats are polluted with unwanted foreigners as well. The European carp, *Cyprinus carpio*, was introduced into Australian waters more than a century ago. These piglike fish went hog wild, especially in the Murray-Darling River Basin (Australia's largest), where they now account for as much as 90% of the total fish biomass and reach densities of one adult per square meter. Through their sheer numbers and rooting behavior, carp degrade river habitats by destroying aquatic plants, by increasing water turbidity and nutrient loads, and by competing with native species for space and food. The river, which humans began to dam and to pollute with agricultural runoff during the last century, was made favorable for carp, and the carp then further debased this aquatic environment.

Around the world, exotic species both large and small are threatening native aquatic ecosystems. Apart from the intentional introductions, one major source of invasion occurs via the ballast waters of transoceanic ships. When these ships empty their holding tanks after an overseas journey, they often disgorge small aquatic lifeforms that may survive to establish new homes. This is how Eurasian zebra mussels, Asian clams, European green crabs, and hordes of other invasive and destructive species arrived in North American waters.

Commercial aquaculture (fish and shellfish farming) now supplies about one-third of all seafood consumed worldwide, but it is another major source of biopollution. Several species of Asian carp, for example, are raised widely in the United States, but they often escape from containment ponds and cause trouble in natural watersheds. Other injurious species invade foreign waters as hitchhikers with farmed animals. A Japanese mollusk that drills into oysters, a South African marine worm that deforms abalone shells, and an aggressive species of Asian eelgrass are among the many nonindigenous predators, parasites, and competitors that hitched rides into U.S. waters via imported shellfish or other products of commercial aquaculture.

The National Research Council, an advisory arm of the U.S. National Academy of Sciences, has ranked exotic species as among the most serious threats (the others being habitat destruction and overfishing) to global biodiversity in aquatic and marine ecosystems. What can be done? One traditional method of biological control has been to introduce parasites or predators from an exotic pest's native homeland. But such transplantations are dangerous because ecological consequences are hard to predict, and the new immigrants might become serious pests too.

Once an exotic species establishes itself in a region where there are no natural enemies, it can be nearly impossible to repress. So, to keep exotics at bay, preventive measures should always be paramount. Tighter policing of ballast-water releases would be helpful, for example. For species reared in aquaculture, another prophylactic route is potentially available, that of making these animals sterile. Several methods are available to render certain species of hatchery fish infertile: surgical castration (clearly applicable only on small scales), the use of sterile hybrid animals (a practice often inimical to goals of high hatchery production and continuity), and the physical induction of sterility by shock treatments.

The latter technique warrants elaboration. When the recently fertilized eggs of some fish species are jolted by specific chemical treatments, pressure, or heat, many of them spontaneously become triploid (i.e., the embryos henceforth carry three sets of chromosomes rather than the normal two). Such fish often develop normally, but when they grow up, they are sterile because their eggs or sperm have odd chromosome numbers. Thus, even if some of these triploid fish were to escape into nature, they could not breed, and any ecological harm they do would be transient. For this reason, some observers think that all organisms reared in commercial aquaculture should be sterilized as standard safety practice. However, where this method has been attempted experimentally, some fraction of the treated fish remain at least partially fertile.

Sterilization of fish by triploidy induction can be considered a form of genetic engineering, but another experimental path, the production of daughterless fish, accomplishes the same end and comes closer to the usual connotation of a GM technology. In the early 1990s, it was discovered that fish, like some other organisms such as fruit flies, carry a gene that when regulated in particular ways during early animal development turns most individuals into males. By genetically tinkering with this "daughterless gene" using recombinant DNA methods, Australian researchers are trying to engineer carp DNA such that the son-favoring gene is highly expressed. In preliminary tests, they have managed to produce experimental zebrafish broods containing mostly male offspring.

The ultimate goal, of course, is to design and release the daughterless GM carp into the Murray-Darling or other biopolluted rivers. Presumably, the resulting carp families will contain mostly sons, the sons will mate with wild females, and after a few generations the carp population will crash from a shortage of females. If all goes well, by the year 2040 the Murray-Darling Rivers may once again be carp-free.

This is long-term planning, but perhaps less distasteful than other control methods such as chemical poisoning. Also, the project would seem to carry few environmental risks. If it fails because the GM genes stop working in the wild population, for example, the researchers merely will have released more carp into an already degraded habitat. If it succeeds, people will have begun to clean up an ecological mess that they created. I remain skeptical about ultimate success in this advertised endeavor, hence the boonmeter rating, but perhaps my intuition will be proved wrong.

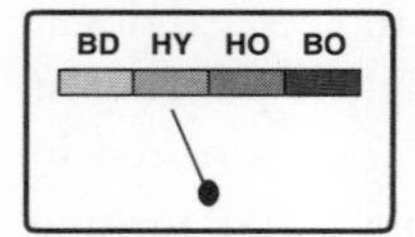

Pesticide Detoxification

The chemical pesticides that humans have developed and deployed over the years sometimes come back to haunt us. A classic case involved the first-ever synthetic insecticide, the organochlorine DDT, that later was banned in the United States. This happened as DDT's devastating ecological impacts and hazards to animal health, including that of humans, became apparent (see chapter 4). The organophosphate insecticides that soon followed, although effective insect killers, had some serious environmental consequences, too. Now, a new generation of scientists is trying to turn the tables once again by genetically engineering ways to detoxify some of the poisonous pesticide residues. This essay highlights two such programs, the first probably a boon and the second more likely a potential boondoggle.

The organochlorine and organophosphate pesticides that accumulate in soils and waters and concentrate in food webs consist of durable, ringlike organic molecules that break down only slowly. Ironically, some insects know how to accelerate the disintegration process. In the last few decades, in response to widespread chemical applications, these resilient insects have evolved enzyme systems that detoxify the poisons by cleaving their ring structures. Scientists in Australia now are hoping to take advantage of these capacities by isolating detoxification genes from pesticide-resistant insects, inserting them into bacteria, and growing the transgenic microbes in industrial fermentation vats to mass produce the enzymes. An alternative (but more difficult) approach might be to mass culture insect cells directly. Either way, these enzymes could then be used to help cleanse dangerous pesticide residues from the surfaces of fruits and vegetables or from polluted waterways, for example. Ecological risks should be low, and, in any event, the program could be curtailed if need be because the cleanup task merely requires the use of enzymes (rather than release of transgenic organisms into the environment).

The second example, also in planning stages, seems egregious by comparison. It involves the pesticide 1080 (sodium monofluoroacetate) now widely used to control exotic mammalian pests introduced into Australia. These include rabbits, foxes, feral cats, feral pigs, and rodents, all of which can do great ecological harm. The typical approach is to set out 1080 as poisonous baits in pest-infested areas. This method has proved quite effective at times and is also recommended because 1080 is a natural plant-based product that readily biodegrades, leaving no lasting harmful residues.

Fluoroacetate, the active ingredient in 1080, is produced as a natural defensive compound by about 38 species of Australian plants, mostly from the genus *Gastrolobium*. As a consequence, over the eons, many marsupial herbivores native to Australia have evolved a certain degree of tolerance to fluoroacetate, a proclivity that is not yet shared by introduced placental mammals; hence the high specificity of the 1080 toxin to the foreigners. However, not all exotic mammals (notably domestic cattle) are considered pests, and therein lies the rub. Every year, a few livestock die from fluoroacetate poisoning, usually after consuming too much native vegetation containing the chemical. So, some scientists now want to protect domestic herds by engineering transgenic ruminant bacteria capable of detoxifying fluoroacetate. The GM microbes will receive the appropriate transgene from soil bacteria (e.g., a *Moraxella* species) and then be delivered to cows to perform detoxification duties in the bovine rumens.

Although the cattle may well gain fluoroacetate protection from these GM bacteria, at least three potential environmental risks, mostly neglected to this point, would seem to be great. First, GM bacteria, once released into cattle, might be difficult or impossible to confine to cows. Feral goats, pigs, or rodents, for example, might pick up the GM bacteria simply by inhabiting pastures housing modified cattle or by feeding on their carcasses. If the GM microbes do jump, such pest species might also gain protection from fluoroacetate, and the whole time-tested approach of pest control by 1080 baits would be imperiled. The total economic costs of lost pest control across the continent would surely far surpass the dollar value of any cattle that die from fluoroacetate ingestion.

Second, the engineered plasmid that will deliver the transgene from soil bacteria to the ruminant bacteria might transfer it to other microbial species, and thereby to other animals as well, with unpredictable ecological consequences. Third, if livestock are imbued with a capacity to handle the toxins that otherwise help protect many native Australian plants from herbivory, the direct and indirect consequences for Australian biodiversity could be overwhelming. Extinction risks probably would increase for many of the native fluoroacetate plant species that already are threatened by virtue of low numbers, and any opening of new grazing areas to fluoroacetate-tolerant cattle would have a cascade of ecological consequences. Already, intense grazing pressures across much of Australia have altered the nature of that continent.

On balance, it appears that alternative, less risky approaches should be examined before any action to release the GM bacteria is considered seriously. Such alternatives do exist. One is to simply live with, or maybe compensate ranchers for, the few cattle that die from fluoroacetate. A second is to keep the cattle and the fluoroacetate plants apart (e.g., by fencing). Ranchers already do this as a matter of course in affected areas. A third approach, which might be labor and cost intensive and will require further research, would be to develop an antidote to fluoroacetate poisoning that could be administered to sick cows as needed.

An even deeper danger lies in the seemingly ill-conceived fluoroacetate venture: It might make the Australian public unduly cynical about the prospects for all forms of genetic engineering. As an antidote to that possibility, the other genetic-engineering initiative in Australia on organophosphate insecticides discussed earlier in this essay should be kept in mind, as well as the ecological and health benefits that could accrue from cleansing the environment of man-made poisons. Thus, the boondoggle rating in the meter applies strictly to the fluoracetate project.

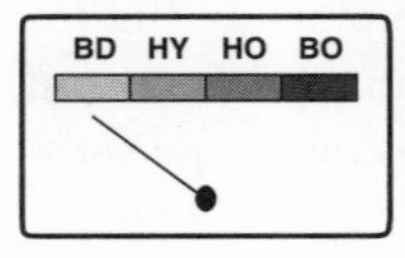

Blue Rose Petals and a Mauve Carnation

Wildflower blossoms are magnificent outcomes of the coevolutionary dances between plants and their avian and insect pollinators, but they also add great beauty and fragrance to human lives. Not entirely content with nature's designs, however, for centuries horticulturists have crossed and artificially selected many plant varieties for improved showiness and bouquet. Still not content, in the last decade a growing cadre of plant scientists has begun to use molecular genetic techniques to compose an array of decorative flowers with even more delightful colors and odors.

The Holy Grail of this new-age horticultural effort is the blue rose. Roses, like carnations and tulips, lack natural blue shades because they have no genetic capacity to synthesize blue pigments, such as delphinidin, that help beautify the petals of many petunias and other flower species. In the early 1990s, Australia's Florigene Company (then Calgene Pacific) decided to rectify this little gardening oversight by Mother Nature. Florigene scientists isolated two petunia genes whose enzyme products (flavenoid hydroxylase and dihydroflavonol reductase) produce delphinidin. When they transferred these genes into rose plants, they got—lo and behold—pink roses. It turns out that delphinidin is like litmus paper, turning blue in the alkaline vacuoles (fluid sacks) inside petunia cells, but remaining pink in the more acidic cellular environment of a rose.

What about carnations, which have a somewhat more alkaline makeup? Florigene scientists repeated the GM experiments with this species, and got mauve carnations. The pale purple-blue flowers were not quite what the researchers had sought, but, making the best of the situation, the company in 1996 began marketing the Moondust carnation, the first of the Florigene Moonglow Series of mauve to blue-violet creations, including Moonshadow, Moonlite, Moonshade, and Moonvista. These were the first transgenic flowers of commercial significance. Some of the initial offerings, in pots, sold worldwide for about $10 per plant (plus postage and handling).

Flower color can be influenced by genetic pathways that impact how pigment compounds are produced and expressed and how they interact with one another and with the plant's intracellular and external environment. By fiddling with the genetic component of these interactions, horticulturists now are engineering new color sensations into a collage of plant varieties, such as petunias displaying novel shades of red, or creative petal wave-patterns. The first ever such GM plant was an orange petunia, created in 1987 by introducing a pigment-producing gene from corn. One of the most intriguing of the more recent GM flowers is the fluorescent daisy, created by inserting a gene for GFP (green-fluorescent protein) from jellyfish. This GM daisy glows bright green under ultraviolet light, and might be a fun novelty at nightclubs or parties.

The cut-flower business is huge, and industry scientists are exploring many additional ways to engineer plants for consumer appeal. For example, Florigene recently patented a gene-based technology for extending the life of cut carnations. Traditionally, florists treat carnations with chemical solutions (such as silver thiosulfate) to prolong the life of blossoms. These act by inhibiting ethylene, a natural plant-produced hormone that triggers aging and petal wilt. In the GM carnations, a comparable outcome was achieved through the genetic suppression of ethylene production within the plants. Florigene touts this creation as an environmentally friendly alternative to conventional chemical treatments.

Plant fragrance is another potentially blooming target for commercial genetic engineering. For example, citrus plants possess an enzyme (limonene synthase) that plays a key role in the synthesis of scent molecules known as monoterpenes. Attempts have been underway to transfer this gene to other species, such as petunias, in the hopes of creating decorator flowers with an appealing lemon fragrance. Collectively, plant species produce hundreds of different monoterpenes and other perfume compounds, so the playing field for this kind of odiferous genetic manipulation is vast.

Other biotech companies, such as NovaFlora, Suntory, Monsanto, and DNA Plant Technology, also have entered the field of flowers. With radiant enthusiasm, they plan to engineer flowering plants and ornamental shrubs for a bouquet of economically relevant characteristics, including:

(a) *Flowering time*. By manipulating genes that give plants a sense of daylight and seasons, the timing and duration of flowering may be altered. NovaFlora already owns the commercial rights to one such gene that changes normal day-length requirements and causes flowers to bloom early. Another research group has engineered asters that bloom in winter as well as in summer.

(b) *Flower arrangement*. Genes influence the distribution of flowers on plant shoots, so this opens many genetic possibilities for rearranging the cut flowers that go into floral arrangements. One such gene, to which NovaFlora owns commercial rights, regulates the growth of apical meristems and inflorescence structure. In some cases, overexpression of this gene might result, for example, in the conversion of single-tip to multi-tip flower heads.

(c) *Height*. Dwarf and compact forms of many ornamental plants sell well. Some of these are created by conventional selective breeding, others through the application of growth-regulating chemicals during development. In many cases, it should be possible to outfit plants with dwarf transgenes, encoding growth-regulating compounds, that achieve similar miniaturization effects without the need for chemical sprays.

(d) *Architecture*. The shapes and growth forms of ornamentals also are of great interest to consumers. Engineering plans are in the works to convert, for example, some bushy plants into taller or climbing varieties, and vice versa.

Genetic tinkering with ornamental plants generally seems harmless enough, and the engineered flowers may help to brighten our lives, hence my positive rating in the boonmeter. I do have one question, though. With all this attention devoted to petunias, carnations, and ornamental species, have researchers forgotten about the blue rose? No. According to company literature, Florigene scientists are still working on it and getting closer.

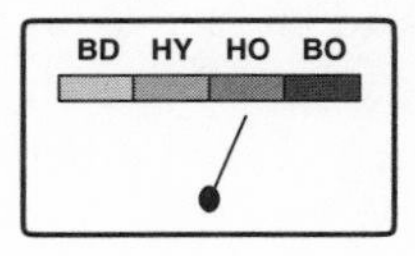

No-mow Lawns

From the thousands of native grasses around the world, approximately a dozen have been adopted by humans for domesticated lawns. Among the most popular are bentgrass (*Agrostis palustris*), a low creeper that forms a dense mat ideal for golf putting greens; Bermudagrass (*Cynodon dactylon*), a major turf species for sport fields, lawns, and parks; Kentucky bluegrass (*Poa pratensis*), which despite its name is native to Europe and Asia; buffalograss (*Buchloe dactyloides*), one of North America's few native turfgrasses; carpetgrass (*Axonopus affinis*), a hardy species well suited for roadsides and airports; centipedegrass (*Eremochloa ophiuroides*), an Asian native that likes sandy soils and warm climates; ryegrass (*Lolium* spp.), which favors cool, moist conditions; St. Augustine (*Stenotaphrum secundatum*), a robust, coarse-textured form used widely for pastures and lawns; tall fescue (*Festuca arundinacea*), a multipurpose grass introduced into the United States from Europe in the early 1800s; and zoysiagrass (*Zoysia* species), a versatile Asian native with high tolerance to drought.

By crossing various forms and using selective breeding, turfgrass scientists have shaped these and related species into more than 100 recognizable genetic varieties with tailored features such as increased shade or drought tolerance, disease resistance, faster or slower growth, wider or narrower leaves, and creeping or upright growth habit. With American homeowners mowing, on average, 40 hours per year, striving endlessly for the perfect lawn, and spending $6 billion annually on lawn care, the stakes are high for grass companies to sprout and market improved varieties. Via traditional plant breeding, a new grass strain takes, on average, about 13 years to develop. Via the new recombinant DNA techniques, genetic engineers hope to speed the process considerably and also invent grass varieties with unprecedented features.

One such trait is herbicide tolerance. Normally, lawn weeds are controlled by spraying selective herbicides, but in addition to the expense and environmental hazards of such chemical use, another drawback is that some weedy species are genetically related to the preferred grasses. For example, annual bluegrass on golf courses is an unwanted and invasive cousin to the desired bentgrass, and this has made it difficult to find herbicides that kill the weed without damaging the turf. Now, as with other plant species such as several food crops (see chapter 4), scientists are using recombinant DNA methods to engineer bentgrass with transgenes for resistance to glyphosate, the active ingredient in many wide-spectrum herbicides. The idea, of course, is that such GM bentgrass could withstand herbicides used to kill bluegrass and other weeds.

Disease resistance is another goal of turf engineering. Lawn grasses are subject to a plethora of pathogen-based diseases such as brown spot, black spot, gray leaf spot, centipede decline, powdery mildew, pythium blight, seedling disorder, rust, stolon rot, bipolaris, fairy rings, and take-all patch. Two such diseases, toward which biotechnologists have directed initial attention, are fusarium blight and dollar spot, caused by the fungal pathogens *Fusarium roseum* (and relatives) and *Sclerotinia homeocarpa*, respectively. For example, GM strains of bentgrass carrying five disease-defiant transgenes have been engineered and field tested for enhanced resistance to dollar spot.

One of the most intriguing goals of ongoing turf research is to engineer "no-mow" grasses with exceptionally slow and stunted growth. This idea gained momentum in 1999 with reports that a gene regulating production of a steroid hormone was manipulated in such a way as to induce dwarfism in experimental tobacco and mustard plants. The plant hormone works much as do related steroids in animals, football players, and weightlifters, normally causing organisms to bulk up. By genetically turning production of the steroid down rather than up, the scientists created miniature plants. The turf company Scotts, among others, is very interested in the prospects for lawn grasses.

Lawnmower manufacturers aren't panicking yet. Even if such GM grass strains can be developed, they presumably will have to gain governmental approval for environmental safety before widespread marketing. One key question, for example, is whether they might cross-pollinate and thereby stunt wild species or standard grass varieties. Furthermore, the primary initial use might be confined to golf courses and new housing developments, as homeowners would be unlikely to tear up and replace existing lawns.

Not everyone is particularly enamored with extensive lawns to begin with, especially in ecologically inappropriate settings (such as deserts). To remain functional and attractive, great expanses of grass require enormous quantities of water, herbicides, and pesticides. Thus, some people think that emphasis should instead be placed upon less intensive landscaping practices, such as a greater use of wildflowers and hardy native plants, where feasible. To these activists, landscaping goals should be modified slightly, from "no-mow" to "no more" lawns.

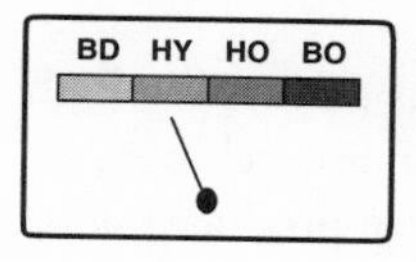

Due to its vast economic potential and popular appeal, lawn engineering has received considerable media attention. To vindicate the hyperbole, grass genetic engineers will have to achieve some major breakthroughs indeed.

Sperm Whale Oils and Jojoba Waxes

Moby-Dick, the gigantic sperm whale sought by Captain Ahab in Herman Melville's 1851 novel, was merely one of its kind under human assault. From relentless hunting across two centuries, numbers of this endangered species (*Physeter catodon*) plummeted from many millions in the world's oceans to fewer than 10,000 individuals today. In the 1980s, the International Whaling Commission (IWC) finally imposed an indefinite moratorium on the commercial hunting of many declining cetacean species, sperm whales included, but some harvesting nations, notably Japan and Norway, still fail to comply forthrightly with all of IWC's stipulations.

The sperm whale is among the largest mammals in the world, capable of reaching 80 feet or more in length, with a weight of 40 tons. However, it was not so much the blubber surrounding the leviathan's body that whalers sought, but rather a fine waxy oil filling a cavernous organ, the spermaceti, inside the whale's prodigious head. The spermaceti, from which the species name derives, holds up to 500 gallons of the precious cream that whalers sold for use as lamp oil, candle wax, watch lubricants, and, more recently, additives for motor oils and transmission fluids, rust-proofing agents, detergents, lubricants for delicate high-altitude instruments, cosmetic bases, and more than 70 pharmaceutical compounds. Beginning in the mid-1800s, kerosene and other petroleum products steadily replaced sperm oil as affordable illuminants and greases, but other uses kept demand high for this exceptional whale substance.

Some 30 years ago, university researchers discovered an amazing analogue of sperm-whale oil in the inch-long seedpods of a desert plant, the evergreen shrub jojoba (*Simmondsia chinensis*). Native to the Sonoran region of the American Southwest (the Latin species name is a misnomer), jojoba produces a fine wax that Indians used for hair care and other purposes, but that now finds manufacturing applications ranging from shampoos and hair conditioners to pharmaceuticals and automotive lubricants. Commercial harvesting of wild jojoba seeds began in the early 1970s, and agricultural plantations quickly were established in semiarid habitats in the United States, Israel, and elsewhere. Jojoba oil is light yellow, highly pure (requiring little refining), and stable to temperatures of 300°C or higher without losing viscosity. In short, it is a fine substitute for whale oil in many applications, and for this reason it alleviated hunting pressures on beleaguered sperm whales.

However, the limited availability of jojoba oil keeps its price quite high relative to petroleum-based products. Accordingly, geneticists are seeking ways to engineer more traditional crop species so that they, too, can produce jojoba oil. Toward that end, experimental mustard plants (*Arabidopsis*) have once again been employed.

The jojoba shrub is unusual among plants in using waxes (rather than triacylglycerol fats) as the storage lipids in its seeds. Wax synthesis occurs in the plant embryos where enzymatic activities convert oleic acid molecules into the very long chain, mono-unsaturated fatty acids and fatty alcohols that make up its waxy substance. Three key enzymes in the process are wax synthase, fatty acyl-CoA reductase, and β-ketoacyl-CoA synthase. DNA sequences for genes specifying these enzymes were identified from jojoba (and *Lunaria annua*, another plant that accumulates long-chain fatty acids in its seeds), cloned into plasmid vectors of *Agrobacterium tumefaciens* (see "Galls and Goals" in the appendix), and inserted into the *Arabidopsis* genome. The enzymes were expressed in these transgenic plants and produced waxy substances that constituted up to 70% by weight of the mustard seeds. The next step will involve attempts to engineer wax-making capabilities into more traditional crop species, such as cauliflower or turnip (related to *Arabidopsis*), for commercial harvests.

Various other crop plants naturally produce diverse fatty acids, oils, and waxes that genetic engineers are aiming to improve for consumer health benefits or other commercial applications. The castor plant (*Ricinus communis*), from which castor oil derives, illustrates both goals. Approximately 90% of castor oil is ricinoleic acid, a useful compound in lubricants, paints, cosmetics, and as a cathartic (bowel purgative), but the seed coat of the bean itself is rich in ricin, an allergenic and potentially fatal protein. Using genetic engineering technologies, scientists are trying to alter the ricin to a less harmful form. They are also trying to modify the plants to produce, instead of ricinoleic acid, a closely related epoxy oil that would be useful in the manufacture of premium oil-based paints.

Because agriculture is involved, this essay could just as well have been placed in chapter 4. I include it here, however, to highlight an environmental issue: the notion that some GM crops producing commercial compounds might on occasion help save wild populations of endangered species. The oils and waxes from GM plants described above, especially had they been available in earlier times, in principle might have helped alleviate hunting and harvesting pressures, respectively, on sperm whales and jojoba plants, species from which those valuable substances otherwise have been derived. Such possibilities, as well as the potential commercial value of such compounds, provide my rationale for a rating of hope in the boonmeter.

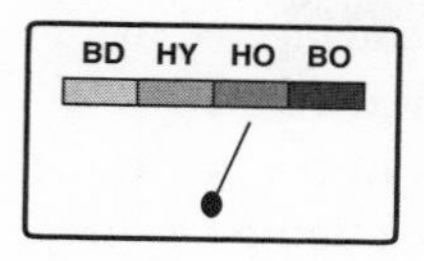

Rescuing Endangered Species

The planet currently is undergoing one of the gravest mass extinctions in its history. According to the World Conservation Union, more than 50% of vertebrate animal species and 10% of all plants are threatened by human activities, and some estimates indicate that about 100 species (many still undescribed) disappear each day, on average. *Homo sapiens* literally is crowding thousands of taxa out of existence. At no previous point in the 4-billion-year history of life has one species exerted such huge influence over so many others.

Recently, a few genetic engineers have begun talking and acting like modern-day Noahs aboard a global ark, striving at the last minute to save at least a few endangered species from the rising flood of humanity. Consider, for example, recent genetic efforts on behalf of the endangered gaur (*Bos gaurus*). This one-ton oxlike bovine, native to India, Southeast Asia, and Indochina, is now on the verge of extinction, having been decimated by overhunting and habitat destruction. The gaur's evident close genetic ties to domestic cattle (*Bos taurus*) made it a logical choice as the first endangered species toward which experimental whole-animal cloning was directed.

In approaches analogous to those used to clone domestic farm animals (see chapter 5), researchers isolated nuclei from skin cells of an adult gaur, injected them into enucleated eggs of domestic cattle, and implanted the resulting embryos for continued gestation in the wombs of surrogate mother cows. Several fetuses developed normally for up to 200 days before they were electively sacrificed for genetic analysis. As expected, the fetuses were pure gaur with respect to nuclear genes, but carried cow-type mitochondrial DNA (from the bovine donor eggs). Results were an initial proof of principle: An endangered mammal (technically, its nuclear genome) could be cloned using enucleated donor eggs and incubation wombs borrowed from a common, related species.

Finally, in January 2001, a cloned gaur was delivered by an ordinary cow. His genes had come from a long-dead adult gaur whose frozen tissue had been preserved at the San Diego Zoo. Named Noah, the gaur calf was the first live-born member of an endangered species to arise from nuclear transfer (NT) cloning. Sadly, Noah died two days later from an illness (clostridial enteritis) that researchers concluded was unrelated to the NT-cloning procedure.

These were the first quasi-successful efforts to clone an endangered mammal by NT, but hardly the first in which researchers employed a foster womb from a common species to incubate the fetus of a rare one. In the 1970s, the principle of interspecies surrogacy was demonstrated when a domestic sheep (*Ovis aries*) gave birth to a mouflon (*O. orientalis*, an endangered species) that researchers had implanted as an embryo into the domestic ewe's womb. Likewise, in the 1980s, an eland antelope (*Taurotragus oryx*, a common African species) gave birth to a bongo antelope (*Tragelaphus euryceros*, a rare species); and, in the 1990s, a house cat (*Felis domesticus*) delivered a desert cat (*F. bieti*, a threatened species). Indeed, a gaur calf likewise has been gestated by a Holstein cow.

For propagating an endangered species, cloning via NT offers one big theoretical advantage over these earlier embryo-transfer methods. Especially when a mammalian species dwindles to low numbers, naturally produced embryos will be in very short supply, even if surrogate mothers of a common, related species are readily available for gestation. In contrast, even a single animal such as the gaur has millions of skin or other cells potentially suitable as nuclear donors for NT cloning. Of course, the technical challenge remains to harvest and use these cells efficiently in delivering cloned animals. Also, unlike conventional transfers of naturally produced embryos into surrogate wombs, the NT cloning procedure merely makes carbon copies of previous genomes in the population. Remove new genetic variety arising from sexual reproduction, and a population's capacity to survive ecological challenges such as disease agents is bound to be compromised.

Can NT cloning (and/or more conventional embryo transfer) contribute materially to the preservation of endangered species? Some scientists are banking on it. In recent years, they have begun assembling genomic banks of cryopreserved cells and embryos from endangered and other species, in the hope that these deposits someday may be used to resurrect declining or even extinct species. In such frozen zoos, tissues are stored indefinitely at extremely low temperatures (e.g., –370°F) and in nitrogen atmospheres for possible future uses as new technologies emerge. Such frozen material sometimes retains remarkable regenerative powers, as illustrated by Jazz, a captive African wildcat (*Felis sylvestris*) born near New Orleans in 1999. As a tiny embryo, Jazz had been experimentally deep-frozen and later thawed and implanted into the womb of a domestic cat. He is living proof that viable kittens can be produced via cross-species gestation of frozen embryos.

Cloning via NT has been achieved in a wide variety of vertebrate animals, including fishes. It has even been accomplished from the cells of animals no longer alive. A case involving a slaughtered cow was mentioned earlier (see chapter 5), and an example involving an endangered species now is available. From cells of two recently deceased mouflons discovered in an Italian pasture, scientists performed the NT procedures and placed the cloned blastocysts into four domestic sheep. Two pregnancies resulted, one of which produced a healthy mouflon lamb now housed at a Sardinian wildlife center.

How long after an animal's death can preserved tissues remain suitable for NT cloning or related genetic engineering manipulations? The blockbuster movie *Jurassic Park* captured the public's imagination for its portrayal of how genetic engineers brought back to life dinosaurs that had been extinct for 150 million years. Are such accomplishments truly plausible? Almost certainly not, because even under the best of postmortem preservation conditions, DNA naturally degrades over thousand-year rather than million-year time scales.

Nonetheless, for species that went extinct recently, resurrection by NT cloning and other genetic engineering techniques is not entirely beyond the realm of possibility, as indicated by a news item that appeared in a June 2002 issue of the journal *Science*. Australian scientists announced plans to resurrect the Tasmanian tiger (*Thylacinus cynocephalus*) from the DNA of three specimens whose tissues were preserved in alcohol more than a century ago. This remarkable Australian species, a marsupial that looked more like a wolf or tiger than a kangaroo, went extinct in 1936 when the last animal died in a Hobart zoo. The *Science* article summarized the new genetic recovery effort: "Opinion varies on the potential success of the venture, with most bets hovering around zero" (p. 1797). The researchers actually involved pegged their own chances at about 5%.

Whole-animal cloning of endangered or extinct species is scientifically exciting and might even help to resuscitate a few endangered species. For reasons of logistics, however, this highly touted approach is unlikely to have major impact on preserving global biodiversity. Furthermore, a serious complaint can be leveled against this approach: It may encourage false senses of hope or complacency that could divert public attention and resources from less glamorous but vastly more meaningful biodiversity efforts, such as habitat preservation. Hence the hyperbole rating on the boonmeter.

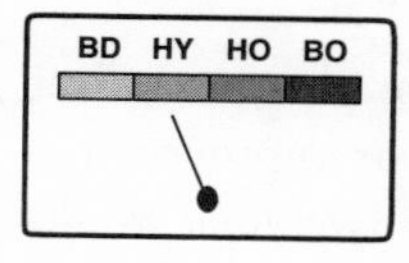

Concluding Thoughts

Genetic engineering in the great outdoors inevitably poses ecological risks along with the potential benefits. The GM creatures are released into the environment intentionally. This contrasts with the usual situation in most other forms of genetic engineering—such as of industrial microbes, barnyard animals, or agricultural crops—where a standard intent is to prevent any unnecessary escape of GM organisms or their transgenes into the wild. In genetic engineering as applied to creatures of fields, forests, and streams, transgenes are purposefully "set free" to do their duty in nature. The question is whether they will properly carry out their intended job without creating additional environmental problems. A recent report by the National Research Council cited unintended environmental impacts as the greatest single category of risks associated with genetic engineering, and among the examples cited were those involving mobile species such as fish.

In some cases, such as modifying plant species for better tolerance to salt or drought, engineering salmon or other native fish for rapid growth, or engineering genetic sterility into pesty feral species, the ecological risks, as well as benefits, can indeed be great. In other cases, such as rescuing endangered species by whole-animal cloning, any ecological concerns from the endeavor itself would seem to be nil. Of course, the magnitude of the dangers and potential boons in any proposed genetic engineering scheme also will vary with the scale of the GM project. For example, to genetically modify whole forests for diminished lignin content inevitably would have far greater ecological consequences than engineering a few fields for trophy blue roses.

Some of the greatest opportunities in outdoor genetic engineering may well relate to environmental remediation, especially of heavily polluted areas such as waste dumps, toxic sites, and despoiled waterways. It would be quite ironic if some new, high-tech GMOs help materially in cleaning up the ecological messes created by earlier high-tech enterprises (i.e., related to mining, fossil-fuel extraction, chemical production, and generation of nuclear power).

The situation is also ironic in another sense. Despite tremendous benefits that in theory could emerge from environmental remediation via bioengineering, science and society have directed precious little attention to this area compared to the vast sums expended in other GM ventures such as crop engineering and the production of recombinant industrial products. The reason for this state of affairs is economics.

Whereas financial incentives for inventing a GM crop or pharmaceutical drug are substantial, who will sponsor environmental safety or improvement? This is a classic example of the "tragedy of the commons": Everyone would benefit from cleaner air and water, but, in a free market system, no person or company stands to profit directly from protecting or remediating publicly owned resources. In the absence of direct market incentives, it often becomes the duty of enlightened governments to contrive economic carrots and sticks for scientific or business ventures so that their actions protect and improve the health of a nation's natural resources. In the case of environmental remediation via GM technologies, this can and should mean increased public funding for research, development, and application.

Even so, there can be no firm guarantees. For example, with regard to pollution cleanup, a recent review by Saylor and Ripp gloomily concluded that the "application of GEMS [genetically engineered microbes] for use in bioremediation has seen little development over the past decade" and "the future use of engineered organisms remains cloudy" (p. 286). That article also noted however, that many of the roadblocks have been economic and regulatory rather than scientific, and that fundamental understandings of microbial biochemistry and genetic manipulation have seen great improvement over that same time. Similar sentiments would apply to plants. Several hundred genetic systems in microbes and many additional ones in plant species are quite well understood now; these in principle could be exploited for environmental monitoring and remediation, given adequate containment and safety assurances.

My feeling is that the environmental sciences (plus particular branches of human medicine, as described in the next chapter) currently offer some of the most interesting but least explored opportunities for further genetic engineering research. If I am right, perhaps technological breakthroughs of broad importance to society will soon emerge from these fields.

7

Genetic Tinkering with Humans

Most of the genetic techniques that might be used to engineer human beings differ little from those that have been used routinely since the 1970s to modify the genes of numerous microbes, plants, and other animals. The ethical issues, however, may be altogether different.

Some of the proposed genetic alterations would apply only to individual recipients of the GM procedures, whereas others would impact the human germline and thereby potentially influence future generations of *Homo sapiens* as well. Should we assume godlike command and attempt to engineer our own species intentionally? Should we take the unprecedented medical and evolutionary reins now available to us and use them to direct our own genetic fates? No prior generation has had the immediate opportunity nor the responsibility to tackle such questions in practical, as well as in philosophic terms. The biotechnology revolution, when applied to the purposeful genetic alteration of people, raises profound issues that will challenge the deepest judgment and wisdom of *Homo sapiens*, the "cognizant" primate.

Gene Therapies on SCIDs

On the morning of September 4, 1990, near a National Institutes of Health Medical Facility in Bethesda, Maryland, two of the most famous figures in the history of human gene therapy met for the first time. French Anderson, a forceful pioneer of human genetic engineering, was introduced to his newest patient, Ashanthi DeSilva, a cherubic four-year-old girl with a grave illness. Ashanthi had been born with SCID (severe combined immunodeficiency) syndrome, a normally fatal genetic disorder that left her immune system unable to fend off infectious diseases. SCID has several genetic etiologies, but in this case the little girl had inherited, from each of her parents, a defective copy of one critical gene on chromosome 20, which left her body unable to produce adenosine deaminase (ADA), a key enzyme in the immune response. So, for her own safety, Ashanthi had been forced to spend most of her young life in quarantine.

On September 5, 1990, for the first of 11 times over the ensuing 2 years, Anderson's team of doctors withdrew some of Ashanthi's blood from which they isolated mature white blood cells (T lymphocytes) for the first-ever experiment in human gene therapy, the purposeful augmentation of a human's genome with artificially introduced genetic material. First, using recombinant DNA techniques, the doctors inserted normal copies of the ADA gene into a disabled viral vector, that in turn delivered copies of the gene to Ashanthi's isolated blood cells. Then, these transgenic lymphocytes were returned to Ashanthi's bloodstream. Over the ensuing months and years, Ashanthi's condition improved dramatically, and she is now a quite healthy young woman, mostly free of the severe effects of her disease.

Nonetheless, this pioneering effort cannot be claimed as a definitive success for human gene therapy because, during the same time period and beyond, Ashanthi also received expensive enzyme replacement therapy in which the ADA enzyme itself, rather than its specifying gene, was delivered to her body. Understandably, the health interests of Ashanthi came first, so all available means were used to save her life. Nonetheless, this medical protocol raised serious scientific doubts about what role (if any) the gene therapy had played in the clinical outcome. Indeed, some studies on ADA-deficient SCID, in other gene-therapy patients, have indicated that few of the genetically corrected white blood cells live long enough or express high enough levels of ADA to provide substantial health benefits to patients. More recently, researchers have identified further regimens that appear to improve ADA delivery via gene therapy and thereby possibly reduce the clinical symptoms of SCID.

The first experiments in clinical gene therapy to achieve unequivocal success, at least temporarily, were reported in 2000. They involved a different form of SCID (type XI) and a slightly different procedure. Hematopoietic stem cells (which are capable of differentiating into all types of blood cells) were removed from bone marrow of two infants and incubated with a retroviral vector carrying a normal copy of the babies' defective gene (for a cytokine receptor of an interleukin protein). Ten months after the transgenic cells were returned to the patients' bones, both babies showed marked immune system improvement that appeared to be attributable to the gene therapy. Soon, several more children with SCID similarly had their immune systems restored by gene therapy.

These outcomes were welcome news for gene therapy because they followed close on the heels of a major setback for the field. On September 17, 1999, the medical community was shocked and saddened by the first human death resulting from a gene therapy trial. Jesse Gelsinger was an 18 year old who had volunteered to undergo gene therapy for a hereditary condition that left his body without the capacity to produce the liver enzyme ornithine transcarbamylase (OTC). Doctors used a GM adenovirus to transfer into Jesse's veins a normal OTC gene, but just four days after the transfection, Jesse died from a massive systemic immune reaction to the supposedly inactivated adenovirus vector. The Food and Drug Administration quickly shut down further clinical trials at the hospital site and blamed the researchers for inadequate oversight. This tragic episode, together with practical and ethical concerns about several gene therapy trials elsewhere, brought this neophyte field of genetic medicine under critical scrutiny by government agencies, outside panels, Congress, and the news media.

Additional setbacks for gene therapy trials occurred in 2002 in France. Two young patients under treatment for SCID both unexpectedly died from leukemia. It turns out that the viral vectors used to deliver the desired transgenes accidentally inserted into and activated the same cancer-causing gene (*LMO2*) in each patient's body. This shocked everyone. A long-standing theoretical concern for the field of gene therapy—that cancers sometimes might arise by insertional mutagenesis when a viral delivery vehicle happens to insert into the genome improperly—suddenly had become tangible.

Nonetheless, in a written scientific commentary following some of the troubling news about gene therapy trials on SCID, French Anderson (2000) wrote: "If one were to believe the news media, gene therapy is both a scientific failure and unsafe. Is this gloomy picture true? Fortunately, no" (p. 627). Anderson went on to suggest that gene therapy, like any major medical revolution (such as organ transplants, antibiotic development, or the introduction of monoclonal antibodies) takes many years to succeed, mature, and become accepted as standard medical practice. Gene therapy, he argued, was still experiencing the growing pains of youth, but "no other area of medicine holds as much promise for providing cures for the many devastating diseases that now ravage humankind" (p. 629).

Gene therapy research on SCIDs has been a mixed blessing for the enterprise of human genetic engineering. On one hand, it entailed pioneering science that opened people's eyes to the tremendous possibilities of the field. On the other hand, it raised grand expectations for improved human health that have yet to be realized, and it has also unmasked serious, unanticipated health dangers of the techniques. Thus, in the boonmeter, I assign this endeavor a neutral ranking at present.

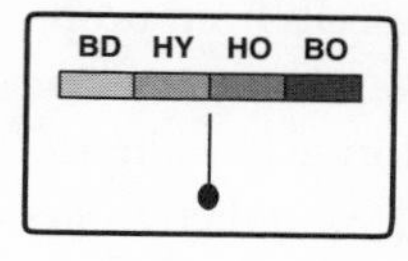

Gene Therapies in the Works

In a sense, gene therapy is nothing fundamentally new, but merely another way to deliver therapeutic substances into people's bodies. Few of us think twice before swallowing aspirin for a headache or taking antibiotics to cure a bacterial infection. These treatments dispense helpful chemicals that our bodies cannot manufacture. Many standard medical practices go a bit further by prescribing enzymes or other functional protein products for individuals with specific genetic deficiencies. For example, blood-clotting factors are routinely administered to hemophiliacs, as is insulin to diabetic patients who are unable to produce adequate levels of this substance on their own. Gene therapy delivers DNA rather than proteins or other substances and thereby shifts the production site for health-enhancing chemicals from outside the body to inside. In gene therapy, if all goes well, a transgene, shuttled to a person via recombinant DNA methods, activates within the patient's cells to produce desired quantities of a useful drug.

Traditional sources of exogenous drugs may be a native medicinal plant, a conventional pharmaceutical factory that extracts or synthesizes valuable compounds from nature, or, in recent years, GM microbes, plants, or barnyard animals (chapters 3–5) engineered specifically to biomanufacture therapeutic compounds for delivery to humans by injection or orally by food or pill. With gene therapy, all such indirect routes are bypassed. Instead, a patient's cells are directly modified so that they produce the desired compound endogenously. The recombinant DNA techniques used to alter human beings are essentially similar to those that genetic engineers use to modify other forms of life.

Scientists have identified specific genes that underlie more than 1000 different hereditary diseases, and the precise molecular basis of many more such genetic disorders undoubtedly will be characterized in the near future. In principle, any of these could be suitable targets for gene therapy, but only a handful have been explored to date in about 450 preliminary gene-therapy trials involving a total of approximately 4000 human subjects. The following are examples of genetic disorders (apart from SCID) under experimental examination in fields of human gene therapy.

Familial hypercholesterolemia (FH), a high-risk factor for heart attacks in young people, stems from genetic mutations that prevent liver cells from properly metabolizing low-density lipoproteins (LDLs). From liver sections taken from several FH patients, cells were isolated, genetically transformed by a retrovirus carrying the normal LDL receptor gene, and reinserted via catheter into the patients' livers. Due in part to variable patient responses, the trials subsequently were discontinued, and no lasting benefit, nor harm, was demonstrated.

Cystic fibrosis (CF), a horrible human disorder that causes respiratory systems to fill with thick mucus, results from mutations in a gene (*CFTR*) that regulates the flow of salt ions across cell membranes lining the lungs. In more than a dozen clinical trials, scientists have used various vectors and protocols in attempts to ferry normal-functioning *CFTR* genes into afflicted patients. Some promising but inconclusive results have been obtained.

Other notable medical disabilities likewise targeted for clinical gene-therapy trials include AIDS, Alzheimer's disease, peripheral and coronary artery disease, rheumatoid arthritis, various cancers, hemophilia, various T cell immunodeficiencies (in addition to SCIDs), multiple sclerosis, muscular dystrophy, and osteodysplasia. Even *in utero* gene therapy (IUGT) is being contemplated because many genetic disorders result in irreversible damage to the embryo or fetus before birth. At the time of this writing, there have been precious few clinical success stories, of any kind, clearly attributable to gene therapy itself, but much extraordinary knowledge is being gained in these endeavors, and many of the researchers involved remain convinced that the technical scientific hurdles will soon be overcome (see next essay).

Any shift from an exogenous to an endogenous source of therapeutic compounds raises pragmatic and philosophical questions. How expensive is gene therapy compared to alternative procedures (now and in the future), and what are the broader ramifications for health-care systems? What will the risks and benefits of gene therapy be compared to traditional treatment options, if any, including standard methods of drug delivery? One important point is that ingested or injected drugs usually degrade quickly in the human body, and thus, must be administered repeatedly. In contrast, in theory (but seldom in practice thus far), successfully inserted transgenes could express proteins constitutively in certain kinds of tissues throughout a person's life. Many diabetics, for example, might be delighted with a one-time gene-therapy treatment that thereafter would free them from the need for daily insulin injections. Nonetheless, circulating levels of drugs delivered by pill or needle might often be far easier to control or reverse than those produced by transgenes implanted in the genome.

Another issue concerns where the transgenes settle. A key research challenge in nearly every case is to design suitable vectors that can deliver a transgene to the target tissue, but not to unintended sites in the body where it might do more harm than good (see next essay). An important special instance is whether a transgene might find its way into germline cells, in which case the patient's descendants could inherit the genetic alteration, for better or for worse. For this reason, it is often useful to distinguish clearly between somatic gene therapy intended to impact only the patient directly and germline engineering that could have extended genetic consequences.

Other challenging issues in somatic gene therapy may be similar to those for standard means of drug administration. For example, to question what constitutes a "defective" gene justifying therapeutic replacement is analogous to questioning what constitutes an "abnormal" physiological state warranting treatment by conventional drugs. By almost everyone's standards, cystic fibrosis would justify treatment, but many less serious conditions will raise far more difficult philosophical issues as to whether gene therapies should be employed.

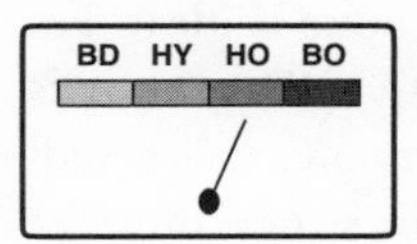

Clearly, tremendous opportunties exist for human gene therapies, but medical researchers will have to achieve further breakthroughs before this hope might transform into health boons for broad segments of society.

New Angles on Gene Therapy Vectors

Why, after almost 15 years since the first clinical trials, hasn't human gene therapy blossomed into the powerful medical revolution that many informed enthusiasts continue to promise? The short answer is that various technical hurdles have proved to be higher than originally supposed. The most serious snag has been the development of suitable vectors that successfully deliver a transgene for proper expression in the targeted cells.

One general approach, illustrated by the first experiments on SCID (see "Gene Therapies on SCIDs"), involves first removing some of the patient's cells for genetic transformation in a test tube or a Petri dish. This procedure permits the physical isolation of specific classes of targeted cells. In the case of Ashanthi DeSilva, mature white blood cells were taken from her body and modified genetically, using a retroviral vector, before being returned to her bloodstream. For a time, the cells appeared to elevate Ashanthi's immune response, but the effect was fleeting and more subtle than hoped, probably due to poor expression of the transgene in the altered cells and to normal cell deaths during blood turnover. Later experiments on SCID type XI had more success, likely in part because undifferentiated stem cells (see following essays), rather than mature lymphocytes, were transformed.

The other general approach is to modify cells directly in the patient's body, without ever removing them. This presents greater challenges in cell targeting but in theory could be a straightforward and flexible procedure, less unpleasant to the patient. Traditionally, viruses have been the favored chauffeurs for carrying transgenes to living human tissue, and adenoviruses and retroviruses are still the most popular vehicles for delivering transgenes in lab experiments and in clinical studies. The suitability of viruses as genetic vectors stems from their small size, relative ease of manipulation, and their evolved ability to hone in on specific human tissues, invade cells, and manipulate cellular machinery to make additional copies of themselves. One of their dangers, forcefully brought home by the unexpected deaths of two patients under gene therapy for SCID, is that the virus might insert adjacent to and activate a cancer-causing gene in the patient.

Several classes of virus are under development as transformation vectors for gene therapy. Each has strengths and weaknesses, but none is ideal. For example, adenoviruses are easy to grow and highly infective. They usually express genes without inserting themselves into the host genome, thus avoiding the cancer-induction risk. However, they also stimulate strong immune reactions that clear the GM virus from the host body, making this vector ineffective for long-term gene therapy. Retroviruses, in contrast, can be effective long term because they provoke limited immune responses; however, they do insert into the host genome, and this raises fear that they might induce cancer-causing or other mutations. Furthermore, retroviruses work only in actively dividing cells. Adeno-associated viruses are invisible to the immune system, readily infect both dividing and nondividing cells, and usually integrate at "safe" locations in the host genome. However, they are hard to grow, and their exceptionally small size limits the length of therapeutic DNA they can shuttle into a cell.

Another risk of using viral vectors became tragically apparent in the gene-therapy trial that killed Jesse Gelsinger. Often, GM viruses can be injected only once or a few times into a patient before they provoke a noticeable immune response. In extreme cases, such as with Jesse, this reaction can be fatal; but even when it is not, it tends to annihilate the viral vector and the infected cells, thereby blocking production of any desirable transgenic protein that the GM virus may have introduced to the patient in gene therapy. For this reason, and especially in the aftermath of the Jesse Gelsinger's death, researchers have sought alternative means of transgene delivery that would fly below the radar of the human immune response.

By the late 1980s, scientists already knew of one such plausible route. In a process that is somewhat like shrink-wrapping a transgene, nucleic acids can be coated with positively charged lipids, thereby creating a "lipoplex" that tends to stick to cell membranes and pop genes through them. Although this method can be effective in introducing foreign DNA or RNA into a mammalian cell, transgenes often fail to activate once inside. In 1989, while working to improve this situation, Philip Felgner, a researcher at a small biotech firm in Southern California, included, as a control in his lipoplex experiments with mice, some animals that were injected only with naked RNA. To everyone's surprise, the control mice, rather than the lipoplex ones, produced the transgenic protein. Thus began a new approach, in some ongoing gene-therapy trials, of injecting naked nucleic acids into human tissues. It does not always work well. Cells other than liver and muscle, for example, are refractory to this form of delivery. However, the direct injection of naked nucleic acids has provided a significant clinical tool in some cases (see next essay). In a cytoplasmic variation on this theme, mitochondrial DNA molecules have been microinjected into mouse cells in experiments to develop gene therapies for deleterious mitochondrial mutations.

To improve the cellular entry of naked nuclear DNA, electroporation sometimes is used. Foreign DNA is injected into target tissue such as skin, muscle, or a cancerous tumor, and electrodes apply a mild electrical field that temporarily punches small holes in the cell membranes. Often, the DNA then migrates through these pores dozens of times more efficiently than without the electrical pulses. Researchers also are pursuing and extending the lipoplex approach, trying different combinations of lipids, polymers, and other molecules that might help cells take up and express foreign DNA. Some of these vector constructs are sufficiently complex so that they act almost like real viruses. They offer a preliminary glimpse of what nonviral vectors of the future may look like as gene therapies move farther into and beyond clinical stages of development.

Most of the transgenes introduced by nonviral vectors have resulted only in short-term gene expression in target tissues, in part because the transgenes have not stitched properly into the genomes of recipient cells. In many cases, transient expression can be medically beneficial. The delivery of vaccines, the treatment of certain cancers, and tissue replacements are examples of potential gene-therapy applications where temporary, as opposed to constitutive, gene expression is desired. However, in various other DNA therapies, where transgenic expression is desired to serve the patient long-term, GM viruses may remain the delivery vehicles of choice. In those cases, the virus must be engineered to avoid the body's immunological radar and to incorporate and properly express itself in nontransient cells.

Basic research into gene-delivery systems has provided significant insights regarding mechanisms and operations of human cells, as well as viruses. These emerging scientific discoveries offer hope that technical hurdles, which have stymied the wide practice of human gene therapy to date, will someday be overcome.

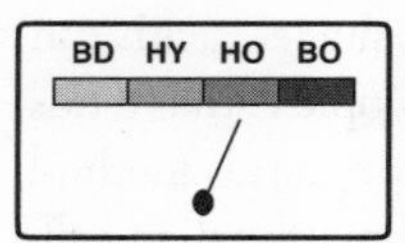

Tissue Therapy via Gene Therapy: The Angiogenesis Story

Angiogenesis is the growth of new blood vessels. It has been the topic of considerable research because insufficient blood flow through diseased or clogged arteries can be responsible for serious medical conditions ranging from heart attacks (the leading cause of death in the United States) and strokes to inadequate perfusion of the extremities, sometimes requiring limb amputations. Recent efforts to stimulate angiogenesis via gene therapy illustrate some of the promises and frustrations of human genetic engineering and introduce the notion that gene therapy can be connected to tissue therapy.

In the human fetus, a protein known as vascular endothelial growth factor (VEGF) plays a role in stimulating normal circulatory development. In 1995, the late cardiologist Jeffrey Isner introduced an experimental gene-therapy treatment for patients with vascular disease. His team coated a tubular catheter with DNA from the gene that encodes VEGF and inserted it into the leg veins of patients with severe peripheral vascular disease. When inflated, the catheter's surface transferred some of the DNA to cells lining the blood vessels. Other trials were conducted on patients with clogged coronary arteries, in which case the VEGF gene sometimes was injected directly into beating heart muscle. These clinical procedures apparently promoted the growth of new arteries from existing ones, occasionally creating detours around blocked vessels. Several dozen patients received the treatments, many of whom afterward reported diminished pain and improved physical endurance.

Since then, a number of angiogenesis trials have explored variations on this theme using different delivery vectors and application techniques. In one approach, the VEGF gene was packaged in an adenovirus that then was injected into the hearts of patients with severe angina. Initial results were promising. Most of the experiments to date have been phase I clinical trials addressing safety issues. If and when these are passed, phase II clinical trials will address actual effectiveness of the therapies. The hope is that such treatments someday may offer an alternative to heart bypass surgery and other traditional operations.

Gene therapies for other tissue types likewise are being pursued in medical research laboratories and in clinical trials. Cancerous tumors offer an interesting example in which the goal is to destroy offensive tissues, rather than to promote the growth of desirable ones. Scientists recently engineered a lipid-coated plasmid so that it carries a gene (*HLA-B7*) specifying a protein that sparks the body's immune response against cancerous cells. This GM plasmid can be injected directly into a tumor. In a phase II trial applying this approach, tumors shrank and life was prolonged for 8 of 73 patients (11%) with aggressive melanomas that had been refractory to other cancer treatments. In a second clinical example, promising initial results followed the delivery of another cancer-fighting gene (interleukin-2) to head and neck tumors unremovable by traditional surgery.

A different sort of gene therapy recently was applied to human buttocks. Skin exposed to the ultraviolet radiation of sunlight is subject to DNA damages that can lead to skin cancer. The problem would be much worse were it not for DNA-repair enzymes that exist naturally in skin cells and mend most molecular injuries. Researchers recently developed a lotion, containing DNA-repair enzymes, that when applied to the skin may augment the body's UV damage control. When tested experimentally on UV-exposed buttocks, the skin cells apparently took up the enzymes and showed marginally improved levels of DNA repair.

To date, most of the gene-therapy successes have been somewhat anecdotal, and it remains to be seen whether the approach will blossom or wither in coming years. As more studies enter phase II clinical trails, an important challenge will be to critically distinguish between real results and placebo effects. Especially when life-threatening or other severe maladies are involved, patients desperately want to believe that medical interventions have provided relief or cure, and they often self-report improvements in their health even in the absence of hard clinical evidence. The power of such positive or wishful thinking was shown recently in another medical context. Surgery for osteoarthritis of the knee is conducted annually on more than 300,000 Americans, with patients often reporting significant improvement after the procedure. However, a large government study concluded, to everyone's surprise, that the surgery *per se* is physically worthless if not harmful. In randomized blind trials, volunteers that had been given sham surgeries actually walked faster and climbed stairs better, after recovery, than those who had gotten the genuine operations.

Placebo effects likewise complicate interpretations of gene therapy trials, especially for procedures like angiogenesis where any tangible improvement in the treated conditions can be difficult to substantiate or quantify. Thus, pending doubleblind experiments that critically address the possibility of placebo effects, reported outcomes of various gene therapies must be interpreted cautiously. Indeed, in one such study, patients who received placebo injections for vascular deficiencies "improved" just as much as those receiving genuine VEGF injections. Of course, for more dangerous or invasive procedures such as heart surgery, any contemplated efforts to conduct placebo experiments will raise daunting ethical as well as procedural issues.

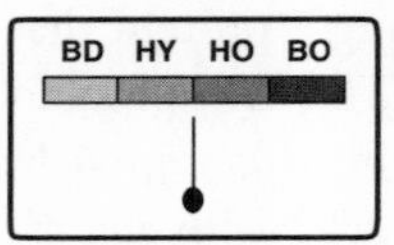

Thus, overall, the jury is still out as to whether tissue therapy via gene therapy will be a boon or a boondoggle for human medicine.

Tissue Therapy via Therapeutic Cloning

More than 40 years have elapsed since Joseph Murray and his colleagues at a Boston hospital successfully transplanted a kidney between identical twins. This landmark approach was later extended by the medical community to other organs (e.g., heart, liver, lung, and pancreas) and to transplants involving more distant relatives and unrelated individuals. Transplants between unrelated individuals are especially challenging because, unless ameliorative actions are taken, the immune system of a transplant recipient sooner or later rejects the alien cells. To alleviate this problem, donor and recipient typically are matched as closely as possible for genes underlying immune responses, and immunosuppressive drugs also are administered. Such procedures are fairly common and have saved many lives. Nonetheless, modern transplantation surgery remains risky due to inherent immunological intolerances of patients to foreign tissue.

Thus, many research professionals are excited about "therapeutic cloning," a new GM approach that in theory should avoid the immunorejection problem. In this procedure, genes in cells to be transplanted originate from the patient, who therefore serves in effect as both donor and recipient. Because the donor and recipient tissues have identical genotypes, presumably the immune system would not recognize the implanted tissue as alien. Another reason for enthusiasm about therapeutic cloning is that this approach gives scientists welcome opportunities for basic research on human genetic disorders as they unfold during cell and tissue development.

The notion of therapeutic cloning for tissue or organ reconstruction in humans traces to the development of nuclear-transfer cloning methods for sheep and other farm animals (see chapter 5). As applied to human cells, the procedure might work as follows: A suitable cell is removed from a patient and its nucleus is inserted physically into an enucleated egg. The egg then begins to multiply in a test tube, and, from the developing mass, pluripotent cells (those that possess a capacity to differentiate into multiple tissue types) are induced to grow replacement cells needed by the patient. Nerve cells might be grown to treat Alzheimer's disease or spinal cord injuries, skin cells could be used to repair burn damage, retinal cells for macular degeneration, pancreatic cells for diabetes, hematopoietic cells for leukemia, neuroglia cells for multiple sclerosis, and so on. When returned to the patient's body, the cloned cells in such tissues or organs ideally would repair or replace the damaged body part, without evoking immunological rejections.

Several technical challenges must be overcome before this approach is medically viable. First, NT techniques developed for farm animals will have to be improved and adapted to our species. Second, cells in the proliferating mass must be generated in such a way that they indeed are pluripotent at the outset. Third, the developmental potential of those flexible cells then must be channeled to produce the specialized kind of tissue that the patient requires. Fourth, methods must be devised to put those now-dedicated cells together properly to make a therapeutically useful tissue or organ. This may take place naturally when the cells are placed in a patient's body, or in some cases it may be accomplished initially *in vitro*. For example, replacement skin tissue for burn victims might be constructed by seeding the cloned cells onto sheets of a polymeric scaffolding substance. Finally, tissue therapy must be conducted such that the cloned cells do no harm when returned to the patient. It would be disastrous, for example, if even a few cells in the transplanted tissue began to divide in an unregulated, cancerous fashion.

Of course, ethical issues will have to be addressed as well. When the initial oocyte created by NT begins to divide into two cells, then four, then eight, and so on, when does the cloned mass become a new human being worthy of protection under the law? Opponents of therapeutic cloning often contend that an individual arises at the exact moment that the first cell appears, such that any sacrifice of an early cell mass, even for medical purposes, is tantamount to slaughter. Proponents of therapeutic cloning view this notion as nonsense. How, they ask, can a few amorphous cells be granted legal rights that take precedence over those of sentient human beings in desperate need of cell therapy? Remarkably, in U.S. society, most of the debate over the possible legalization of therapeutic cloning hinges on this one emotion-laden philosophical issue.

In such public discussions, a common error (or often, an intentional argumentative ploy) is to equate therapeutic cloning with reproductive cloning (see "Whole-Human Clones"). Although the initial laboratory steps in the two procedures are identical—both begin by inserting a cell nucleus into an unfertilized egg—that is where the similarity ends. In reproductive cloning, the GM egg would be reimplanted in the womb and allowed to grow into a fetus and baby, the intent being to generate a fully functional and independent human being genetically identical to its predecessor. In therapeutic cloning, the early clump of preimplantation cells that comes from the GM egg would be grown *in vitro* and used to produce replacement tissues for medical rehabilitation.

After open deliberations, if a democratic society decides to ban therapeutic cloning on moral grounds, so be it. However, it would be a great shame if that decision arose merely from semantic confusion between therapeutic and reproductive cloning. For this reason, and because the word cloning itself carries such emotional baggage, some people have suggested that "therapeutic cloning" be replaced by less provocative phrases such as nuclear transplantation, tissue rejuvenation, or NT cell therapy. Indeed, if the word cloning had never been used by researchers in the context of NT cell therapy, quite likely this medical procedure would not have catapulted so forcefully into the public and political spotlight.

The artificial cloning of cells for therapeutic purposes offers some of the greatest possibilities in all of genetic medicine for saving human lives and improving the health of countless citizens. Thus, I find it horribly sad, as well as ironic, that this promising new approach has been met with strident social and political opposition in the United States. Tragically, opponents of therapeutic cloning, who would elevate the rights of cells above those of sentient human beings, might possibly cause our nation to squander this marvelous opportunity for improving human health.

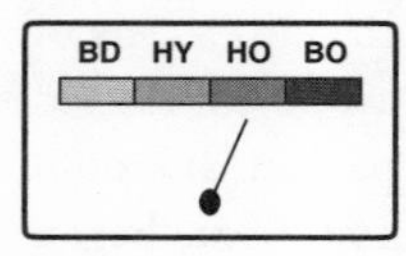

Embryonic Stem Cells

During normal sexual reproduction, the nucleus of a fertilized egg or zygote contains two nearly matched sets of genetic material, one from each of the two parents. This never-before-seen mixture of genes will interact to direct an individual's biological development, from pre-embryo to the person's death perhaps decades later. However, in the first two to four rounds of cell division, most of the RNA and protein molecules that orchestrate initial development are critical holdovers that were produced and deposited into the egg cytoplasm (the envelope surrounding the nucleus) by the mother's genome. Much of the primary activation in the zygotic genome occurs after a ball of about 4–16 pre-embryonic cells has formed.

After further cell-division rounds (each roughly doubling the total number of cells), a fluid-filled cavity begins to form within the growing cell-ball. This hollow sphere, known as a blastula or blastocyst, consists initially of about 100 cells. To one side of its central cavity is a knob of undifferentiated cells, the "inner cell mass," from which embryonic stem (ES) cells are derived. They eventually will give rise to all of the 260 or so cell types that make up a person's tissues, organs, and body structures.

The entire blastocyst, still floating free in the mother's reproductive tract, finally implants into the female's uterus during the second week after fertilization. Then the differentiation of the embryo's body parts begins in earnest. For example, a heart takes shape and begins beating by week four. One week later, the embryo reaches a length of about one-third of an inch. By week eight, rudimentary precursors to all body structures in the adult are present, and the developing individual is termed a fetus. By the end of the first trimester (12 weeks), the two-inch long fetus becomes recognizably human (as opposed to another animal). By the end of the second trimester (24 weeks), the fetus is almost a foot long, and at about 40 weeks a baby is born.

The initial happenings are basically the same under therapeutic cloning via nuclear transfer (preceding essay), except that in therapeutic cloning the nuclear genome of the zygote is a replica of the nuclear genome of the donor somatic cell, rather than a novel amalgamation of genes from an egg and a sperm. Under therapeutic cloning, however, development of the early cell mass occurs *in vitro* and is truncated at the preimplantation stage, when ES cells are harvested and grown for medical purposes. It is the pluripotency of ES cells, their ability to differentiate into multiple cell and tissue types, that makes these cells valuable as wellsprings for tissue rejuvenation. In principle, ES cells for research or medical purposes can also be obtained from naturally produced pre-embryos or embryos. Indeed, such surplus specimens are discarded routinely at fertility clinics conducting *in vitro* fertilizations. Regardless of how ES cells are obtained, once isolated they can be grown in laboratory cell culture more or less indefinitely. However, stem cell lines do tend to become less functional and less viable over time.

When exposed to particular environmental stimuli *in vivo* or *in vitro*, ES cells spontaneously differentiate into any of a variety of cell types, such as muscle, pancreas, skin, or liver. Scientists have identified several ways to manipulate ES cells grown in culture so that they develop into particular lineage types. For example, applying retinoic acid tends to channel ES cells to become neurons; serum albumin and other factors can stimulate ES cells to differentiate into blood cells; and "protein 4" induces ES cells to form mesenchymal cells that when naturally present in the living body produce connective tissues or transport vessels of the circulatory and lymphatic systems. However, far more research will be necessary (see next essay) before we learn how such ES cells might be harnessed to medically relevant tasks.

About 4 million Americans have Alzheimer's disease, 1.5 million have Parkinson's, and 200,000 suffer with serious spinal cord injuries. These are just a few of the individuals and medical conditions that might be early candidates for therapeutic treatments using ES cells to regenerate damaged tissues. But in the United States, opposition to stem cell research has been stiff in some circles. President George W. Bush, for example, has stated that he is opposed to cloning of any sort, and Leon Kass, whom Bush appointed to chair a National Bioethics Commission (to advise the president on biomedical issues and be "the conscience of our country") is also a gut-level opponent. Recently, that commission produced a mixed vote on whether therapeutic cloning should be banned. An advisory panel of the National Academy of Sciences, the most august scientific organization in America, issued a 2001 report affirming support for therapeutic cloning (while also calling for a ban on reproductive cloning). Polls suggest that the U.S. public is also strongly in favor of ES cell research for medical purposes.

In a decision with huge sociopolitical as well as scientific ramifications, President Bush then decreed that federal funds not be used for research on ES cells other than those from an estimated 64 cell lines that purportedly were available in August 2001 (the usable number actually turned out to be far lower, perhaps ten or fewer). Many researchers in the United States view this ban as a serious impediment to medical advances because of insufficient genetic variety as well as limited availability of government-approved stem cell lines. The construction of additional cell lines has since been taken up without federal support, for example, at Stanford University.

Some observers hailed the 2001 Bush decision as a reasonable compromise between two competing ethical concerns (i.e., that harvesting ES cells is equivalent to embryonic murder, and that a complete ban on ES research would deny millions of disabled children and adults the therapeutic potential of tissue rejuvenation). Others interpreted the Bush decision as political pander to religious conservatives, an action that will seriously impede health care in the United States and drive key medical research overseas. Saudi Arabia, China, England, and Malaysia are among the countries actively considering therapeutic cloning of ES cells as a part of their own medical biotechnology initiatives.

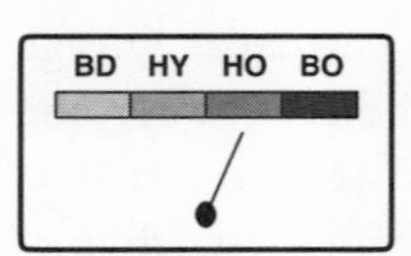

I have already stated (previous essay) where I stand on these issues. My hope is that ethically responsible research into therapeutic cloning, if otherwise left unfettered, will help lead human medicine into a much brighter future.

More on Stem Cells

The legislative and media attention that has been directed toward stem cells is rather surprising given the still rudimentary state of understanding about these pluripotent cells. Nonetheless, stem cell research has lurched forward on several fronts. Here I describe several available examples, emphasizing those that are most amazing, provocative, and controversial.

If therapeutic cloning is someday to become a mainstream procedure for regenerating damaged tissues and body parts, a severe problem to be overcome is the shortage of available human eggs in which to incubate, in cell culture, the nuclei from donor somatic cells. One possible solution might be to transfer the nuclei from human cells into the enucleated eggs of species other than *Homo sapiens*. A bold step in this sci-fi scenario recently became reality, when Chinese scientists revealed that they had shuttled nuclei from human cells into cultured enucleated rabbit eggs. At the Sun Yat-Sen Medical University in Guangzhou, researchers removed the chromosomes from rabbit eggs and replaced them with nuclei taken from skin cells of a seven-year-old boy. Approximately 100 of the nuclear transfers reportedly succeeded, and one of the dividing cell masses in culture developed for three days, almost to the blastula stage.

This was not the first time a human nucleus has been ferried into another mammal. Years earlier, Jose Cibelli and his team of therapeutic cloners at the biotech firm Advanced Cell Technology (ACT) in Worcester, Massachusetts, successfully replaced nuclei in cow eggs with those from human white blood cells and skin cells lining the mouth. The rationale was to explore the feasibility of cow oocytes as incubator sites for therapeutic cell cloning. Bovine eggs are available in abundance; those used for the ACT experiments came from ovaries collected in slaughterhouses. Some of the nuclear transfers did indeed lead to human-cow blastocysts (human nucleus, bovine cytoplasm), but the results were poor and inconsistent overall.

Another challenge to be overcome in therapeutic cloning involves tumorigenicity. When pluripotent ES cells are injected experimentally into mice, they sometimes grow cancerously, producing invasive, debilitating teratomas (tumors made of different types of tissues). The same would probably occur in humans. Even differentiated tissues grown from ES cells in culture pose this danger because they can remain contaminated by at least a few undifferentiated, potentially cancerous ES cells. One response to this threat would be to engineer ES cells that self-signal their contaminating presence. Researchers recently accomplished this feat by hooking a jellyfish gene for green fluorescent protein (see "Reporter Genes" in the appendix) to a gene that is expressed only in human ES cells. Thus, in such engineered lines, any unwanted ES cells that remain in a transgenic tissue culture would light up with a green warning glow.

In principle, an actual solution to the tumorigenicity problem is to engineer fail-safe suicide mechanisms that would cause transplanted ES cells to self-destruct if they became tumorigenic. A properly designed transgene could be engineered into the donor genome before the NT procedure. Later, in response to a specific hormone or other chemical administered to the patient, that transgene would activate and kill any runaway cell in which it was housed. This experimental approach already has been developed and employed to destroy renegade ES cells grown *in vitro*, in mixed cell culture.

Another approach under preliminary exploration provides an alternative to conventional notions of therapeutic NT cloning. Parthenogenesis or virgin birth (see introduction to chapter 5) is a natural process in some species, by which an unreduced egg cell (one with a diploid rather than a haploid number of chromosomes) is stimulated to develop into a new animal without the benefit of fertilization by sperm. In 2002, researchers at ACT reported the derivation of primate ES cells via parthenogenesis. They isolated normal precursor egg cells from macaque monkeys, *Macaca fascicularis*, and used chemical treatments (rather than sperm) to activate these diploid cells to begin dividing. Several of the unfertilized eggs developed to the blastocyst stage and contained harvestable ES cells. By sidestepping the natural sexual process in the creation of primate embryos, this bizarre procedure opens yet another potential pathway to generate ES cells for therapeutic applications.

Embryos are not the only sources of stem cells. Most differentiated tissues in children and adults also contain small pools of undifferentiated cells ("adult stem cells") that can proliferate extensively to produce all of the cell types normally found in that tissue, and sometimes others. For example, hematopoietic stem cells in the bone marrow continually produce the various types of blood cells; neural stem cells in the hippocampus and olfactory bulb of the brain can generate neurons and their supporting neuroglial cells; and some adult stem cells from muscle reportedly can produce blood cells as well as muscle cells. The refreshment of tissues via cell turnover and replacement is a normal part of a body's metabolism.

At least one company (StemSource) is trying to capitalize on the regenerative potential of adult stem cells. It offers liposuction patients a deal. For a mere $1500, the company will sift through the patient's shed fat for potentially useful stem cells and store them for five years. The promise is that if stem cell research someday pays off, these cells could be used to produce genetically matched replacement tissues for the foresighted donors. It might even become possible to modify these or any other stem cells, or the therapeutic procedures themselves, so that the cell lines become useful as more universal donors. So far, however, all this is little more than wishful thinking, and no clinically proven applications yet exist for such stem cells.

Adult mammals appear to contain at least 20 major types of somatic stem cells, but active scientific debates currently surround the latitude and effectiveness of those cells in refurbishing different types of tissues, both *in vitro* and *in vivo*. To the extent that adult stem cells someday may prove effective in human medicine, they could replace ES cells for such purposes and thereby bypass all ethical objections against the use of embryonic cells. On the other hand, many medical researchers are concerned that compared to ES cells, adult stem cells will prove to be a blind or narrow alley with only limited clinical promise. Certainly, these adult cells seem not to have the full pluripotent scope of ES cells.

Societies, as well as researchers, eventually will have to judge which of the many lines of inquiry into stem cells are ethically responsible as well as genuinely fruitful. Of course, opinions are likely to vary from person to person, but the final judgments should at least be scientifically informed.

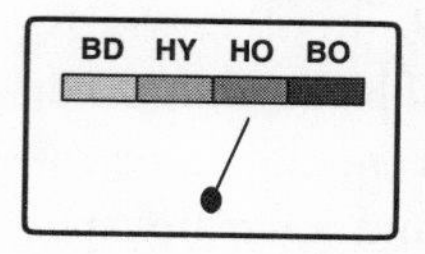

Whole-Human Clones

Virtually all basic research to date on therapeutic cloning via nuclear transfer has used mice, farm animals, primates, or other mammals as experimental surrogates for humans. Recently, that situation has begun to change, despite ethical concerns about extending such investigations to our species.

In early 2001, advertisements appeared in Boston-area outlets for women between the ages of 24 and 32 willing to donate eggs that would provide a platform for cloning experiments. The researchers at ACT who placed the ads sought women who were physically and psychologically healthy and who, for the current purpose, were not interested in producing a child for themselves or for adoption. Instead, any human embryos produced in the therapeutic-cloning trials would be used to isolate ES cells. Twelve suitable young women stepped forward, as did a number of other individuals who volunteered to donate small skin biopsies or a few tiny ovarian cells to provide the nuclear genomes for the NT procedures.

The first such cloning attempts at ACT took place in June 2001, using NT methods entirely analogous to those that produced Dolly the sheep and many other barnyard animals (chapter 5). In all, 71 eggs were manipulated before the first pre-embryo successfully arose. However, it survived only to the six-cell stage and hence did not generate an inner cell mass from which ES cells could be harvested. Undaunted, the ACT researchers also tried to generate early human embryos by parthenogenesis. Using techniques similar to those used to clone macaque monkeys to the blastocyst stage (previous essay), they exposed 22 unreduced (diploid) human eggs to chemicals that altered the ionic composition of the cells and stimulated their proliferation. In cell culture, six of the eggs developed into what appeared to be blastocysts, but none clearly presented an inner cell mass containing the desired ES cells.

The ACT researchers first published their results in the online *Journal of Reproductive Medicine*, and the report created quite an uproar, in part because of claims that these cell masses were the first cloned humans. Three esteemed members of the journal's editorial board later resigned, ostensibly in protest that the data failed to support the authors' conclusions (the genetic identity of the cloned cells was not verified by hard scientific criteria, for example).

Undoubtedly, someone will take the next technical step and permit a cloned human embryo to come to term in a woman's womb. In other words, the first human baby derived from reproductive cloning will be born. Indeed, shortly after Dolly the cloned sheep was created, scientists (notably a reproductive physiologist, Panos Zavos, in the United States, and a fertility specialist, Severino Antinori, in Italy) announced to the news media their intent to clone human babies. So, too, did at least one private company (Clonaid). In each case, the stated intent was to help infertile couples produce children from their own body cells. Soon, media reports began to appear of cloned babies having been born, but these early cases probably were hoaxes. In one sense, it hardly matters; regardless of the validity of existing claims that human clones have been artificially created and born live, the fact remains that the outcome is plausible, and the feat almost certainly will be verifiably accomplished soon.

Any endeavor to generate cloned babies runs counter to strong public sentiment and against the edicts of many agencies and governments around the world. Many people view human cloning as ethically repugnant, going against nature's or God's design by undermining our deep-rooted values of individuality, uniqueness, and personal worth. Human cloning often has been portrayed as a means for rich or narcissistic individuals to make copies of themselves or for unscrupulous dictators to produce brigades of identical jackbooted soldiers. Most animal-cloning experts are revolted by the prospects of human reproductive cloning, and also point out medical dangers associated with the process. As emphasized repeatedly by Ian Wilmut, inventor of farm-animal cloning, 276 failures accompanied the production of Dolly the sheep, and washout rates at least this high can be expected in any initial human-cloning effort. Opponents of reproductive cloning contend that because each aborted outcome leaves behind a "murdered" zygote, blastocyst, embryo, or fetus, it is a moral indictment against humanity.

How, then, do proponents of human reproductive cloning defend such efforts? They contend that infertility is a medical disability and their goal is to help infertile people have children of their own. Brigitte Boisselier, the director of Clonaid, argues that the opportunity to reproduce, and choosing how to do so, are basic human rights. She emphasizes that cloning may be the only reproductive option available to infertile individuals or couples and that society's ethical concerns are misplaced because a cloned baby is merely a "belated twin" of its predecessor, fundamentally no different from monozygotic twins that commonly arise during normal sexual reproduction. Proponents of reproductive cloning also suggest that with diligent research, technical hurdles can be overcome such that, in time, cloning procedures will become medically safe.

Thus, advocates view human cloning as merely another assisted reproductive technology (ART) to be welcomed into the arsenal of medical approaches that overcome common infertility problems. Approximately 12% of young couples suffer from at least a modest shortfall in fertility from such causes as sperm shortage, sperm immotility, blocked fallopian tubes, abnormal uterine anatomy, shortage of eggs, incompatibility between eggs and sperm, and so on. Over the years, several ARTs have been introduced and incorporated as standard practice into mainstream medicine, often after initial societal resistance on ethical grounds.

For example, artificial insemination is an ART that became popular in the 1970s as a way of circumventing physical or ejaculation difficulties in men, or to allow single or gay women to have children from anonymous donors (a practice made legal in the United States by 1978). The development of cryopreservation methods (freeze-preserving gametic cells without killing them) made artificial insemination even more convenient. The next major ART, *in vitro* fertilization (IVF), got rolling in 1978 with the birth of the first test-tube baby, Louise Joy Brown. This technique, in which an egg and a sperm are united in a test tube and the resulting embryo is implanted in a woman's uterus, can overcome various infertility problems, such as sperm immotility in the father or structural damage to the reproductive tract of the egg-donating biological mother (if the IVF embryo is incubated in the womb of a surrogate mother). By 1995, nearly 150,000 IVF babies had been born worldwide, and more than three times that number soon may be present in the United States alone.

Each of these and other ARTs caused tremendous controversy when introduced, but the ethical objections eventually diminished as the benefits and safety of the procedures became clearer. It will be interesting to see whether the same course of events attends human reproductive cloning via nuclear transfer or parthenogenesis. Will these techniques also become routine someday, or will they remain so morally objectionable or dangerous as to be forever banned?

Reproductive human cloning has garnered sensationalized attention by the media. Nonetheless, the efforts have not proceeded very far, and in my

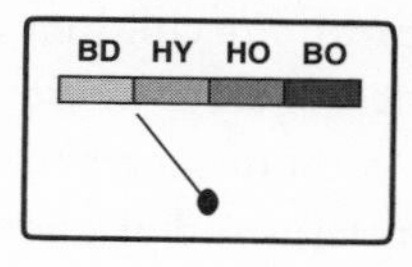

opinion they also tend to have much less compelling (albeit sometimes justifiable) rationales than therapeutic cloning. Thus, human reproductive cloning to date merits an overall rating merely in the hyperbole/boondoggle range.

Engineering the Germline

All 100 trillion cells in each adult human trace back to a single diploid cell, the fertilized egg. Nearly all of the cell divisions are mitotic, faithfully replicating the zygotic genome into armies of derivative cells. Most of these are somatic cells that constitute heart, skin, nerves, liver, muscle, blood, and the body's many other parts. However, a significant event occurs early in embryogenesis when a few thousand amoebalike cells begin to form the primordial gonadal tissues of the developing embryo. It is from this small band of dividing cells (founders of the germline), that all of the individual's eggs or sperm eventually will arise, and it is only these germline gametic cells whose DNA has a chance to survive the inevitable death of the individual (i.e., by contributing to zygotes of the next generation).

Whatever hopes or ethical concerns may attend the practice of human somatic gene therapy, they are likely to become magnified in deliberations on germline engineering. In germline engineering, the genetic alterations can affect not only the person receiving the GM procedure, but also his or her descendants. Germline DNA is potentially immortal. Each piece of human DNA in existence today traces its history through an unbroken chain of ancestry extending back across thousands and thousands of generations, each link in the chain forged by gametic cells. Looking ahead, all people in the future will derive their DNA from the germlines of those of us alive today. Thus, any tinkering with the human germline is not to be initiated cavalierly.

Yet, in the first half of the twentieth century, various eugenics movements in effect did just that by encouraging reproduction by people with supposedly desirable genetic features (e.g., fair skin, square jaws, or favored character traits) and discouraging reproduction by those deemed undesirable. A nadir for eugenic efforts came in Nazi Germany, going so far as to promote racial extermination purportedly to improve humanity's gene pool. Now, in the twenty-first century, a new brand of eugenics has become imaginable: direct alteration of the human germline via recombinant DNA techniques. However, the misplaced goals and horrible actions of last century's eugenics crusades should give serious pause to any modern-day genetic engineer who might feel an urge to direct our species toward some genetic ideal or imagined utopia.

On the other hand, would not we wish, if possible, to rid our children and grandchildren of the fear of horrendous genetic disorders such as cystic fibrosis or Huntington's disease, much as we have labored without regret to eradicate external agents of human disease such as the smallpox virus and plague bacterium? Hopefully, this is much closer to the kind of genetic engineering possibilities that societies will address from ethical as well as practical standpoints in any new eugenics movements.

Moral issues aside for the moment, what kinds of technical methods from the field of recombinant DNA might be available to engineer genes in the human germline? In principle, three direct genetic approaches can be envisioned: alter zygotes, alter germline cells within the body, or alter germline cells *in vitro* before returning them to the human population. These are discussed in turn.

Because the germline, as well as the somatic cells, originate from a single diploid cell via mitotic divisions, any genetic modification of the zygote normally will be passed intact to the germline. In chapter 5, various methods were described by which zygotes from barnyard animals are engineered to carry particular transgenes. For example, a foreign piece of DNA can be microinjected into the egg directly, or it can be ferried in by a GM viral vector. Comparable approaches can be envisioned for humans. However, the animal studies also indicate that much can go wrong. Only small fractions of tampered zygotes typically incorporate the desired transgene successfully, relatively few of those survive *in utero*, and relatively few of those survivors develop into fully healthy animals. Likewise, genetic modifications in humans may carry unexpected health consequences even in the best of outcomes.

The second plausible approach is to deliver foreign DNA to the germline within the patient, for example, by using a disabled viral vector that targets the appropriate cells. A likely outcome of this method is genetic mosaicism in which some germ cells carry the transgene and others do not. All else being equal, the earlier and more effectively the foreign DNA is delivered during an individual's development, the higher the fraction of transgenic gametes he or she will produce. A potential drawback of this approach is that each experiment is conducted within a live individual, who therefore could bear the health burden of any procedural failure or unintended action of the transgene.

The third approach is to remove gametes or other germline cells from the patient, insert a specific transgene into them in a test tube, and screen the resulting cells *in vitro* for viability and presence of foreign DNA. The transformed cells then might be returned to the individual's reproductive tract, or, more likely, used immediately to produce a new embryo (e.g., via artificial insemination or by IVF). The technical feasibility of such germline manipulation recently was addressed in another primate, a rhesus monkey named ANDi (from "inserted DNA" spelled backward). True to his name, ANDi carries a foreign gene for a green fluorescent protein (GFP) delivered to the unfertilized egg from which he arose. Scientists isolated 224 mature oocytes from adult rhesus females, exposed them to GM retroviruses housing the GFP gene, artificially fertilized the manipulated egg cells with rhesus sperm, and implanted the fertilized eggs into the monkeys' reproductive tracts. Three baby monkeys eventually were born, but ANDi alone proved to carry the foreign DNA.

A human female produces only about 400 mature egg cells during her lifetime, releasing them one by one during her monthly ovulatory cycle. By contrast, males produce sperm in abundance. A typical ejaculate contains about 600 million spermatozoa, and a man's lifetime production may exceed 10 trillion cells, roughly 1000 sperm for each and every heartbeat. Genetic engineers are exploring ways to capitalize upon this ready abundance of male sex cells.

In experimental rodents, scientists isolated germline stem cells from healthy males and transplanted them into the testes of infertile males, thereby restoring fertility in the latter. Similar testis-cell transplantations have been initiated in clinical trials on humans, where the intent is to recover fertility in cancer patients that have undergone irradiation or chemotherapy. An extension of this approach, already attempted in mice but not in humans, is to correct additional genetic defects during the transplantation procedure. GM viral vectors, for example, have been used to introduce transgenes to isolated germ cells. By implanting germline cells with specific gain-of-function or loss-of-function genes, it may be possible to correct serious genetic disorders or otherwise improve the health of any resulting progeny.

Notwithstanding recent progress in understanding some of the basic genetic techniques that will be required for any contemplated efforts in human germline engineering, the field has not yet moved far. Thus, it remains for now within the realm of hyperbole, both positive and negative, as people ponder what good and bad might emerge in any new-age eugenics.

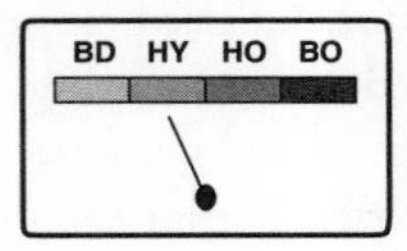

More New-Age Eugenics

A different eugenics approach, if humanity should choose to employ it, could emerge from new biotechnologies that permit the molecular diagnoses of genetic diseases. Molecular diagnosis is the application of DNA-level assays to reveal the presence or absence of a specific heritable disorder in an individual, and molecular screening is the extension of genetic diagnosis to large numbers of genes or people. If data from genetic screening were employed on a massive scale to influence or direct people's reproductive decisions, in principle the human gene pool could be altered appreciably.

Even before completion of the Human Genome Project in 2001, resulting in a draft sequence of nearly all 3 billion nucleotide pairs in the nucleus of a human cell, the medical profession had developed molecular-level assays that could detect particular disease-causing variations on the normal versions of a wide variety of genes. Clinical tests were made available for genetic abnormalities such as AAT deficiency, cystic fibrosis, fragile X syndrome, hemophilia, Lesch-Nyhan syndrome, Marfan syndrome, muscular dystrophy, neurofibromatosis, phenylketonuria, sickle-cell anemia, thalassemia, and many others, and the list will grow dramatically as more genes are studied and characterized at molecular levels.

A classic example of genetic diagnosis is that for Huntington's disease (HD), a fatal genetic disorder whose symptoms, usually beginning in midlife, involve uncontrollable movements of the body and progressive dementia. More than 25,000 people in the United States alone suffer from HD, and nearly 100,000 more may develop HD later in life. This disease results from mutations in a gene located at the distal tip of chromosome 4. In 1993, after two decades of arduous detective work, researchers unveiled a detailed profile of the genetic defect. The molecular signature of HD is an expansion of a CAG nucleotide repeat unit, from 10–30 copies in the nonmutated gene, to several dozen more copies in disease-causing versions. This breakthrough discovery led to a diagnostic laboratory test for presence or absence of deleterious forms of the gene. By running tiny samples of an individual's relevant genetic material through an electrophoretic gel that separates molecules by size, any flawed version of the gene reveals itself as an abnormally large piece of DNA.

Now, anyone who undergoes this genetic diagnosis and is found to carry a defective copy of the gene will know with high probability that HD will strike in midlife. Not everyone wants such information. Even in families where HD is present, individuals at risk often opt not to take the diagnostic test. Others prefer to preview their fate, to dispel the mental anguish of not knowing. Further, some people elect to take the test to assist in their own reproductive decisions. Without knowing the test outcome, a prospective parent at risk for HD (one who has a parent or sibling already diagnosed with the disease) has a 25% chance of transmitting an HD gene to any child. With test results available, however, those odds move either to 50% or 0% depending, respectively, on whether the molecular diagnosis was positive or negative for presence of HD mutations.

In principle, if every copy of the defective HD gene in the population were identified and blocked from transmission to children, Huntington's disease could be eliminated immediately. Such is the theoretical power of molecular genetic diagnosis coupled with reproductive actions. But who will make such reproductive decisions? It is one thing if the informed judgment on whether and when to have children is left to prospective parents, but quite another if the decision is made by interested third parties such as the medical profession, insurance companies, and/or governments.

Technological improvements in genetic diagnoses sometimes can aggravate the ethical issues. For many years, amniocentesis (in which doctors retrieve fetal cells sloughed into the amniotic fluid) and chorionic villi sampling (of cells from embryos as young as eight weeks) have been used to help screen for particular genetic conditions *in utero*. In a recent extension of this approach known as preimplantation genetics, DNA is isolated and amplified by PCR (polymerase chain reaction; see "Test-tube Gene Cloning" in the appendix) from blastocysts or even preblastocysts, and then tested for genetic defects. Suppose a pre-embryo, embryo, or early fetus is found to have Down's syndrome, or a defective copy of the HD gene, or some other highly serious but sublethal genetic disorder. Should it be aborted, and, again, who decides: the parents, insurance providers, right-to-life activists, or governments?

In at least one case, massive molecular genetic screening already has been used, in conjunction with reproductive actions, to effect a decline in the incidence of a serious genetic disorder. Tay-Sachs (TS) is a genetic disease of severe mental retardation, paralysis, and death, usually before the age of five. Only homozygous individuals (those with two copies of the TS mutation) have this incurable illness; heterozygotes (persons with only one copy of the mutation) display no overt symptoms, but nonetheless are carriers, liable to pass the defective gene to their children. Thus, a child of two heterozygous parents can inherit the disease. In the early 1980s, more than 300,000 Jewish volunteers worldwide submitted to a TS test, and many took the results into account in subsequent marriage and family plans. The program was criticized by some on ethical grounds, but it did result, for example, in the virtual elimination of Tay-Sachs disease from the population of Hasidic Jews in Brooklyn.

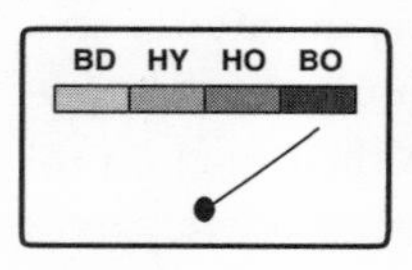

In summary, medical diagnostic procedures often emerge as practical and valuable by-products of molecular investigations into the genetic bases of human hereditary disorders.

Concluding Thoughts

When considering the significant efforts, research dollars, and media attention devoted to human gene therapies and the negligible clinical successes of such enterprises to date, one cannot help but wonder whether these resources might better have been focused on other healthcare initiatives, such as disease prevention or environmental improvement. For example, man-made toxins, carcinogens, allergens, and other pollutants cause countless more illnesses and deaths than will be remedied by any foreseeable gene therapies applied to human genetic disorders in the near future. Over the past two decades, if society's primary health goal had been to improve the lives of the greatest number of people, vastly more effective avenues have certainly been available.

Of course, gene therapy and other forms of genetic engineering are not necessarily in competition with standard health-care approaches, nor is clinical success to date the only measure by which research on genetic engineering should be evaluated. Increases in our basic understanding of human genetics, plus improved prospects for future clinical successes, must be factored into the equations, too. Nonetheless, it is true that gene therapy and human genetic engineering have been huge health disappointments to date, relative to initial high hopes and expectations. Only if that situation changes will human genetic engineering fundamentally alter, during our lifetimes, the practice of medicine as we have known it.

Staunch opponents of human genetic engineering often invoke the metaphor of the slippery slope, down which societies will inexorably slide if even a small first step is taken. For example, they demand legislative bans on therapeutic cloning from fear that such practices would irresistably lead to human reproductive cloning, or they argue against genetic manipulations of somatic cells (e.g., gene therapy procedures) because these would inevitably lead to dreaded tinkerings with the human germline. Although the slippery-slope metaphor is logically unsound, it remains popular, seductive, and powerful, not least because it implicitly conveys an impression that the starting point for this anti-GM advocacy position rests on high moral ground.

If such metaphors are to be used at all, I prefer another version: the terraced slope. In this image, societies would use scientific evidence and reason as bases for enlightened decisions regarding human genetic engineering and would consciously choose to move toward terraces on the mountain deemed to be of net societal benefit (while giving due respect and safety to opposing stances). All societal decisions would continually be subject to revision in the light of empirical experience, and the desirable terraces to be settled upon perhaps adjusted accordingly. Furthermore, this metaphor makes no implicit assumption about the starting point (i.e., the present human condition). In other words, where we are now in terms of gross human happiness and well-being, relative to what might be achieved through genetic engineering, might be near the top of the mountain, but it also might be near the bottom, or anywhere in between. The terraced-slope metaphor places a high premium on societal flexibility, open mindedness, and reasoned judgment. Thus, any decisions made under its image will be facilitated by extensive input from a well-educated citizenry.

Epilogue

"Genetic engineering" can be a misleading term if taken to imply undue analogies to physical engineering. When geneticists tinker with the genetic makeup of plants, animals, or microbes, the outcomes can be far less predictable than when a mechanical engineer builds a house, a bridge, or a dam. Each individual organism changes continually during its development, thus posing an ever-shifting physiological milieu for genetic manipulation. At the population level also, nothing remains static, but rather evolves, often in partial response to any genetic alterations themselves. So, too, will other species evolve with whom GMOs interact. The net results are ever-changing and varied responses to genetic manipulations and levels of temporal unpredictability that far surpass what physical engineers encounter.

Suppose, for example, that flowing water could evolve resistance to any dam meant to impede its progress. Perhaps it could mutate to a more acidic form that quickly would eat holes in the structure, or maybe evolve an enhanced capacity to hurdle over barriers. Engineers would then have to monitor dams continually and have contingency plans for water's many possible evolutionary responses to its new environmental challenge. Due to possible catastrophic dam failure from water's dam-mediated evolution, wise societies would insist on great caution in building such structures, especially upstream of cities or other critical sites. In the real world, mechanical engineers can ignore such hypothetical evolutionary responses by the inorganic entities they seek to control. Biological engineers have no such luxury. Evolutionary changes in living populations occur in response to genetic engineering. This much is guaranteed.

The first three decades of genetic engineering often have been characterized by inappropriate linear thinking in several regards. First, near-term profits have driven much of the commercial enterprise, with limited regard for diffuse or longer term societal and environmental ramifications of genetic manipulations. Indeed, some companies have shown a remarkable disregard for public sensibilities, and a general hubris has hurt their own cause.

Second, a common tendency has been to suppose that a given transgene will affect only the specific organismal trait in question, such as insecticide resistance, or the capacity to produce a pharmaceutical drug. But living creatures are complex beings with multitudinous metabolic and physiological demands underlying successful survival and reproduction. Biological intuition predicts, and empirical experience confirms, that specific genetic traits seldom can be altered without precipitating collateral effects. Cascades of ancillary or compensatory impacts on the transgenic organisms often become apparent only after the fact. The phenomenon of yield drag in transgenic crops (see chapter 4) is one familiar example. Side-impacts of inserting growth-hormone genes into fish (see chapter 6) is another. The prospect of unanticipated side effects must also be borne in mind by anyone contemplating the genetic modification of humans.

Third, a frequent tendency has been to underemphasize, if not neglect, the extended consequences of releasing GMOs into environments. Just as a transgene inserted into a genome can have unforeseen effects on the GM organism itself, a transgenic strain introduced into the environment can have unintended impacts on the ecology and evolution of interacting species. Due to such community-level responses, plausible outcomes and contingency plans should be considered a priori, as well as monitored after a GM strain has been unleashed into nature. Seldom have either been done with full vigor and effectiveness, despite various regulatory procedures in some countries.

A great irony is that many of genetic engineering's greatest promises lie in correcting problems created by the cutting-edge technologies of earlier eras. For example, GM microbes and plants someday may be used widely to decontaminate toxic waste sites, rejuvenate tainted soils, and cleanse polluted waterways of poisonous chemical effluents from manufacturing. Another example is the use of GMOs, or the compounds they produce, to combat emerging disease agents in agriculture and medicine, many of which evolved in response to the unwise applications of chemical insecticides and antibiotics in prior decades. Furthermore, various GMOs will increasingly be used to biomanufacture environmentally friendly alternatives to traditional industrial substances such as poorly degradable plastics and detergents. The new genetic technologies of cloning and other forms of reproductive assistance may even help to rescue (at least in zoos) a few of the thousands of species endangered by human activities.

Despite nearly three decades of experience with recombinant DNA techniques, the ultimate contribution of genetic engineering to the broader human enterprise remains uncertain. I suspect that the field will display some significant parallels with the way nuclear physics developed in the last century. Initially, the prospect of unlimited power through nuclear fission was a powerful allure, and vast sums were invested in relevant technologies. Today, nuclear-power reactors dot the land and supply a modest fraction of the nation's and the world's energy demands. However, the approach has not lived up to its initial hype, and also has created some huge problems, related to safety, security, and contamination, with which societies must deal.

Likewise, various products of genetic engineering will be incorporated into society and certainly will make important contributions to human well-being. Yes, there will probably be some significant local accidents (analogous to Chernobyl and Three Mile Island), in which GMOs harm people, other organisms, or environments. Yes, there will probably be some broader or longer lasting problems of biological cleanup (analogous to the disposal of radioactive wastes), as some GMOs released into the environment create ecological difficulties. Yes, we will have to live with ongoing threats of biotic catastrophes (analogous to nuclear holocausts), especially if recombinant DNA technologies fall into the wrong hands and are used to create biological weapons of mass destruction.

The promise of genetic engineering also differs from that of nuclear physics in important ways. Whereas the primary commercial use of nuclear energy lies in power generation, genetic engineering can find application in a vast range of activities, from agriculture and manufacturing to medicine and environmental remediation. Furthermore, whereas nuclear fission is inherently dangerous to life, many, but certainly not all, of the contemplated uses of genetic engineering are quite safe and benign. Thus, over the longer term, genetic engineering may yet blossom in additional directions that will bring huge and lasting benefits to mankind.

My overarching goal in writing this book has been to convey the sheer wonder of a novel human technological capability, drawn from the raw genetic materials and molecular tools of nature's own evolutionary design. Humankind's newly found capacity to purposefully manipulate genes is surely among the most extraordinary achievements in the history of our species. In some important respects, it represents a fundamental breach, a demarcational crossing, from all that had gone before. This technical achievement, and its ethical as well as pragmatic implications, are surely worthy of our utmost, continuous attention.

Most of the genetic research programs described in this book are still in their infancy and no doubt will soon be expanded or amended considerably (some even before these words appear in print). Various genetic engineering endeavors are proceeding at such breakneck speed that no single person can keep abreast of all salient developments. Nonetheless, because the stakes are high, societies must try to assimilate the new information, contemplate meanings, frame and conduct intelligent debates, and in general take active and enlightened roles in shaping genetic engineering enterprises. I hope that this book will have been constructive in these regards.

Appendix: Tools and Workshops of Genetic Engineering

In this appendix I describe several common laboratory techniques used by genetic engineers, mostly courtesy of unicellular microbes and their evolved expertise in manipulating DNA.

Restricted Activities

A typical bacterium is a single-celled organism measuring about 2.0 micrometer (μm) in length. A micrometer is one-millionth of a meter, so more than 10,000 bacteria lined in a row would barely stretch an inch. Individually, bacteria are too tiny to be seen by human eyes without the aid of a microscope. But from the perspective of an even tinier entity, a viral particle, a bacterium is a behemoth. For some diminutive viruses, 50 specimens placed end to end would barely span the diameter of one bacterial cell, and millions could inhabit the head of a pin.

All viruses are obligate intracellular parasites, spending most of their life cycle within the cell of a host species and using that cell's molecular apparatus to reproduce. Viruses that infect bacteria, as opposed to those that inhabit animal or plant cells, are called bacteriophages, or phages for short. Some, such as the T4 phage that infects the bacterium *Escherichia coli*, look remarkably like lunar landing crafts, each with a polyhedral protein capsule, housing viral DNA, situated above a sheath that resembles a heat shield, all perched atop splayed tail fibers reminiscent of a spaceship's landing gear. When a T4 module alights on the surface of a bacterial cell, it recognizes and attaches to specific receptor sites. Like a plunging hypodermic syringe, the phage then physically contracts and injects its DNA, and only its DNA, into the host.

In one possible outcome, known as the lysogenic cycle, viral DNA next integrates into the host chromosome, and, with the host DNA, replicates and multiplies with each round of host bacterial cell division. The result is a descendent population of bacterial cells all infected with viral DNA, perhaps burdened by the parasite, but still alive. An alternative outcome is the lytic cycle, wherein a phage uses the host's metabolic machinery to make full copies of itself, including protein capsule, sheath, and tail fibers. The phage proliferates wildly inside the cell until the bacterium swells and bursts, like a ruptured balloon. Sprayed out, the new phage particles then infect other bacteria.

As is generally true of host–parasite interactions, bacteria are not always helpless victims in this process. Under the tutelage of natural selection across eons, many bacteria have learned to fight back with defensive weapons that can severely restrict a phage's activities. Bacteria do not have complex immune systems like vertebrate animals, but they have evolved functional analogues, restriction enzymes, that do a comparable job in fighting off infectious viral agents.

For example, *Eco*RI, named after *Escherichia coli*, the bacterium that makes this restriction enzyme, cleaves (digests) DNA wherever it encounters the exact nucleotide sequence GAATTC. Another restriction enzyme, *Bam*HI, produced by the bacterium *Bacillus ambofaciens*, cuts DNA molecules at each GGATCC site. Each bacterial strain can generate and store a specific restriction enzyme because it also has evolved a protective chemical "modification" system that prevents that restriction enzyme from digesting the bacterium's own genome, or else the microbe would self-destruct.

Bacteria produce and use restriction enzymes in part to destroy invasive viral DNA. Via restriction enzymes, bacteria sometimes can turn the tables on invasive parasitic phages, rendering the attacker the attacked. As a consequence, any phage that happens to be naturally resistant to the restriction enzymes of its host will tend to be favored by natural selection, as are any bacteria that evolve more effective defenses against viral invasion. Thus transpires a continuing evolutionary contest of punch and counterpunch between each bacterial strain and its phages, every bout critically refereed by natural selection. One net result of these long-standing evolutionary matches has been the elaboration of a collective diversity of restriction enzymes in various bacterial species. Scientists have identified, isolated, and characterized hundreds of these proteins.

When the first restriction enzymes were discovered and characterized in 1968, their utility in genetic microsurgery was immediately appreciated. Via standard laboratory methods, DNA can be extracted and purified from any species, but initially it comes in unwieldy strands many millions of nucleotide pairs (or base pairs) long. Researchers had yearned for some precise and repeatable method by which DNA could be chopped into shorter, yet well-defined, manageable pieces. Restriction enzymes, isolated from various bacterial species, were just what the Ph.D.s ordered.

Consider, for example, the restriction enzyme *Eco*RI that cuts DNA at the six-basepair recognition site, GAATTC. If we assume for simplicity that A, T, C, and G are equally frequent and randomly ordered in a long piece of native DNA, then *Eco*RI will digest that DNA into smaller fragments typically just a few thousand nucleotides in length. In other words, it will cut that DNA on average about once every $4^6 \cong 5000$ bases. Another restriction enzyme that cuts at a four-base-pair recognition site will produce DNA fragments averaging just a few hundred (i.e., $4^4 \cong 250$) nucleotides long. Thus, bacterial restriction enzymes offer a handy set of precision cutting instruments for the genetic engineer.

Cut and Paste

Modern word processors greatly simplify manuscript editing, enabling writers to reorder phrases with just a few swipes of the mouse and a touch of computer buttons. Two useful editing commands in any word processor are the "cut" and "paste" options. Since the mid-1970s (coinciding with the era of word processors), biologists have had at their disposal analogous cut-and-paste options for editing genomic texts of any species. Genetic engineers use restriction enzymes to cut long pieces of DNA from any biological source into manageable fragments. Geneticists also have a paste function by which they can weld such DNA pieces back together again, often mixing and matching them in new orders.

Earlier, some details about restriction enzymes were glossed over that now are relevant to our discussion. Recall that *Eco*RI clips DNA at each nucleotide sequence GAATTC, which is really just shorthand notation. *Eco*RI actually snips duplex (double-stranded) DNA, and it does so exactly between the G and the A on each of the two complementary nucleotide strands. In the *Eco*RI case, the second strand of the recognition site, read from left to right, is CTTAAG, exactly the reverse of GAATTC. This is a common feature of restriction-enzyme sites (i.e., most are short, inverted repeats [palindromes] of nucleotides on the two tandem strands of template DNA).

Thus, upon close inspection, incisions that *Eco*RI makes are slightly staggered through each duplex DNA molecule, rather than perpendicular across a restriction site:

G * A A T T C
C T T A A * G

where asterisks indicate actual sites of the cut. So, the wound is ragged, with single-stranded pieces of DNA left dangling from the severed strands. In each case, the protruding single strand from one DNA strand is AATT (as read from left to right), and the protruding strand from the other is TTAA.

The fact that the cut is staggered in this way is important because the dangling termini themselves are complementary in nucleotide sequence. This means that the tips of any DNA fragments produced by digestion with a particular restriction enzyme tend to attract and grab hold of one another, like Velcro (see "Hybridizing DNA Molecules"). Such DNA fragments are said to have sticky ends. Velcro is an insecure adhesive, however, and, likewise, the DNA fragments are not yet tightly bonded, an operation that requires one further biochemical step.

Ligases are enzymes that catalyze formation of tight bonds between pairs of suitable DNA fragments, joining them end to end. Like glues or wood screws used to solidify tongue-and-groove joints between dovetailed pieces of wood, ligases complete each joining of juxtaposed DNA fragments. They do so by sealing phosphodiester bonds in the sugar-phosphate backbone of DNA molecules.

Ligases are produced naturally by all organisms as a part of their DNA replication apparatus. In 1967, the first ligase to be isolated and studied came from *Escherichia coli*. Commercial laboratories now routinely isolate and synthesize ligases, as they do restriction enzymes, from various bacterial species and market these to research scientists for genetic engineering or other DNA-manipulating purposes.

Copy and Duplicate

Two other useful functions on word processors are "copy" and "duplicate," which permit the easy replication of short passages and entire computer files, respectively. Beginning in the early 1970s, genetic engineers developed analogous laboratory tools by which they can reproduce portions or complete copies of nature's hereditary texts. Collectively, these "carbon-copying" techniques are referred to as cloning methods.

Gene cloning can be introduced by reference to bacterial plasmids, which are tiny, circular pieces of DNA that reside within bacterial cells. These genetic elements typically lie separate from the main bacterial chromosome and replicate independently of it. Some plasmids also incorporate into the host chromosome, at which time they are called episomes. Plasmids, which range in number from a few per cell to as many as 100, readily transfer from one bacterium to another when the microbes "mate" (conjugate). Despite their small size, many plasmids carry genes for antibiotic resistance or other traits that can be critical to bacterial survival in certain environments.

By virtue of their small size and abundance in bacterial cells, plasmids have become favored tools for genetic engineers for producing clonal copies of a gene from any biological source. First, these circular molecules are isolated from a suitable bacterium such as *Escherichia coli*, and, in a test tube, sliced open by a particular restriction enzyme chosen because it recognizes only one site within each plasmid. The linearized plasmids then are incubated with DNA fragments from a donor source that the biotechnologist wants to clone. These fragments have been generated by digesting suitable DNA from any desired species (e.g., a human) with that same restriction enzyme. If all goes well, the sticky ends (see "Hybridizing DNA Molecules") of the donor DNA fragment sometimes wed with the complementary sticky ends of an open plasmid, recircularizing the genetic construct. The addition of a ligase enzyme then seals the deal, consummating the molecular marriage.

The plasmid, now carrying a piece of foreign DNA, is reintroduced into its bacterial host. There the transgene is replicated, together with the plasmid DNA, as the population of bacterial cells divides and multiplies. In other words, while naturally making copies of itself, each recombinant bacterium also generates clonal copies of the transgene. In some cases, the genetically transformed bacterium may be the end of the story—the final desired repository for the transgene. For example, a genetic engineer's goal might be to use GM bacterial colonies to produce large quantities of a commercially valuable protein, such as human insulin, encoded by that transgene. In other cases, the bacterium may be an intermediary vessel, a biological vector for transferring the transgene into a third-party species. For example, to engineer crops with new metabolic capabilities, agricultural geneticists routinely use soil-inhabiting bacteria to transmit alien genes into various plants (see "Galls and Goals").

Many observers informally date the birth of genetic engineering via gene splicing to a seminal 1973 paper by Stanley Cohen and colleagues. In a test tube, they artificially constructed a hybrid plasmid by joining restriction fragments generated from two native plasmid forms. When reinserted into the bacterium *Escherichia coli*, the recombinant plasmid was shown to replicate properly and to confer upon recipient cells the anticipated combined functional properties of the two parent plasmids.

Unlike phages, which selfishly parasitize bacterial cells, plasmids often pay their rent in various ways, for example, by carrying genes that facilitate genetic recombination in their hosts or by housing genes for resistance to particular antibiotics. Normally, bacteria die when exposed to an antibiotic such as penicillin, ampicillin, or tetracycline. However, any bacterial strain that acquires antibiotic-resistant plasmids may survive and even thrive in these otherwise poisonous environments. Although such plasmid-mediated resistance can be a microbe's salvation, it can be disastrous for humans; many bacterial strains, in hospitals for example, have evolved the ability to withstand even the most powerful antibiotics developed over the years by pharmaceutical companies. For example, more than 90% of all isolates of *Staphylococcus aureus*, one of the most common disease-causing microbes in humans, are now resistant to penicillin.

However, as a screening device for identifying genetically transformed cells, the antibiotic-resistance properties of plasmids are useful to genetic engineers. The screen works as follows: After the GM plasmids have been reintroduced to their bacterial hosts, a scientist can sample from the resulting colonies of cells and test for those that are (versus are not) resistant to the relevant antibiotic. Only bacterial strains that prove to have this antibiotic resistance are likely to have received the transgene.

Suppose now that the human DNA, used to begin procedures, had been an uncharacterized ensemble of restriction fragments, each a different piece of DNA from some unknown location in the genome. Thus, each transgenic bacterium (of which there could be many thousands) likely would carry a different fragment of human DNA. Such a collection of recombinant bacteria composes a genomic library. By various technical procedures, scientists then can sort through this microbial repository of human DNA sequences to find and retrieve cloned copies of a specific human gene of interest. Genomic libraries for many species now are available commercially or from individual researchers and scientific organizations.

Viral Vectors

In genetic engineering, a bacterial plasmid is an example of a cloning vector or transformation vector for a transgene, like a miniature Trojan horse that can enter the otherwise walled (or membraned) fortress that is a cell, and deposit and replicate a foreign piece of DNA therein. Plasmids (and phages, another popular cloning vehicle) are used widely by biotechnologists as vectors to introduce foreign DNA into bacteria. Similar kinds of biological delivery systems have been developed to introduce exogenous genes directly into multicellular organisms. Viruses that infect the cells of plants and animals are prime examples.

Retroviruses are especially useful as transformation vectors, notably in mammals. These single-stranded RNA viruses reproduce by "reverse transcribing" their tiny genomes into DNA molecules that then insert into their host's chromosomes. Once integrated, the viral DNA is transcribed by the host's own molecular machinery into more retroviral RNA genomes, as well as into mRNA molecules responsible for the synthesis of viral proteins. Properly packaged inside these proteins, the newly replicated retroviral genomes then escape the host cell to infect other cells. Via this same process, any piece of alien genetic material that has been inserted into the retroviral genome is likely to be carried along for the hereditary ride, likewise copied and spread through populations of host cells. Technical differences aside, this entire cloning operation roughly parallels the way that plasmids mediate insertion and proliferation of a transgene in a bacterial population.

Viruses come in many varieties and with exquisite host specificities. In genetic engineering, this can be important in helping to confine the spread of a transgene to a given target species or set of related taxa. Many viruses also specialize in specific organs or tissues within the host. Adenoviruses, for example, infect cells of the human eye, respiratory tract, and immune tissue (adenoids), whereas herpes viruses primarily infect cells of the nervous system and skin. This type of host-cell specificity can be useful to genetic engineers when a particular tissue is the intended target for gene therapy. For example, a transgene whose products might alleviate the symptoms of Parkinson's disease could be conveyed to a patient by a virus that colonizes neurological cells. Indeed, this approach was used recently in experimental mice and monkeys, where a transgene, delivered by an engineered virus, induced the animals' brains to produce dopamine, a natural substance whose absence is associated with Parkinson's syndrome. In another example, an adeno-associated virus carrying a human dystrophin gene, when injected into muscle tissues of experimental mice, helped relieve the animals of symptoms similar to human muscular dystrophy.

Another inherent hallmark of viruses, however, is their proliferative, invasive nature, not only within, but sometimes between, species. Such invasions can be hugely consequential: witness the recent colonization of *Homo sapiens* by an immunodeficiency virus (HIV, the retrovirus underlying AIDS) that may have originally jumped into our species from wild African primates. A growing body of scientific evidence suggests that during the history of life, numerous host genes have been transferred even between distantly related species. The frequency and modes by which such DNA commerce has taken place across the leaky boundaries between multicellular species are currently under intense debate, but viruses and other kinds of mobile elements (next essay) are leading suspects as mediating agents.

These observations raise two opposing considerations relevant to current GM practices of using viruses as transformation vectors. On the one hand, viral-mediated "horizontal genetic transmission" (movement of genes between species other than via hybridization) probably has occurred naturally for eons, thus supporting the contention that this biotic process, when now conducted under biotechnologists' auspices in a laboratory, is nothing fundamentally new. On the other hand, viral vectors carrying transgenes may not always confine themselves to the target species, thus countering any contention that safety guidelines are unnecessary in the field release of virus-carried transgenes.

Further, viruses may not always confine their movements to intended cells within an organism. One such example came to light in 2001 when, during a human gene-therapy experiment (chapter 7) using an adeno-associated vector, signs of the virus appeared not only in a patient's blood, as intended, but also in his semen. Thus, a therapeutic treatment aimed to help the individual also might have altered his reproductive (germline) cells. This raised an ethical red flag. The Recombinant DNA Advisory Committee, an oversight group at the National Institutes of Health and the Food and Drug Administration, had forbidden any human gene-therapy trials that alter the germline because such genetic modifications could be passed to future generations.

Two additional points are worth emphasizing. First, many if not most viruses are natural agents of disease, in humans ranging from minor colds to influenza, polio, and smallpox. Thus, before enlisting a viruses' services as a transformation vector, researchers usually take steps to disable the virus genetically. Second, many tumor viruses promote cancers in various ways, such as by turning on a cellular oncogene (a gene that normally regulates cell division properly, but when altered can induce uncontrolled cell proliferation). "Insertional mutagenesis" is a general term for how a virus (even a disabled one) sometimes can insert into and trigger a cancer-causing gene within a patient's body. Scientists normally cannot control exactly where a GM virus enters a genome, so a danger always exists that the virus will insert into a critical host gene, disrupt its function, and cause health difficulties or even death of the transgenic organism (see chapter 7).

The bottom line is that properly engineered viruses can be powerful tools for producing GMOs but also that their use as cloning vectors is not to be taken lightly. Ideally, in any such genetic engineering project, scientists and regulatory agencies should weigh the immediate and long-term risks, as well as the potential benefits, of harnessing these tiny vectors.

Jumping Genes

In 1983, Barbara McClintock received a Nobel Prize for her research, during the 1940s and 1950s, that led to the discovery of "jumping genes." Before that time, most genes were thought to be stationary, well-behaved pieces of DNA all lined up on their respective chromosomes like beads on a string. McClintock's work established that many genetic elements are less stationary, sometimes moving from one chromosomal site to another. Furthermore, many of these jumping genes copy themselves during the transposition process such that genetic replicas come to occupy multiple sites.

Such mobile pieces of DNA are also called "transposable elements" (TEs) or "transposons," for which geneticists have awarded cute monikers such as castaway, gypsy, hobo, hopscotch, jockey, mariner, pioneer, stowaway, tourist, and wanderer. Nearly ubiquitous in the biological world, these jumping genes have varied structural motifs. Transposable elements can hop about the genome in two basic ways: a type I element transposes via an RNA intermediate, whereas a type II element simply excises itself from one chromosomal site and inserts at another.

The type I forms are known as retrotransposable elements (RTEs), and they closely resemble retroviruses (preceding essay) in structure and behavior. Indeed, RTEs and retroviruses are closely related evolutionarily, as evidenced, for example, by the fact that both contain a gene for reverse transcriptase, a key enzyme that catalyzes the conversion of RNA to DNA. When a retroelement jumps, the original RTE actually sits still but is transcribed into complementary RNA molecules that later reverse-transcribe into new copies of the original RTE that then insert at other chromosomal sites. So, some RTEs come to be present in many copies dispersed around the genome.

The proliferative nature of many transposable elements has given these little genomic vagabonds the reputation of being "selfish" elements, concerned first and foremost with their own survival and reproduction, rather than the good of the host organism. This strategy has worked well for them, as evidenced by the fact that active TEs and their less-frolicky descendants (who have lost their mobility) make up a large fraction of the genome in most multicellular organisms. When a mobile element makes copies of itself that disperse across the chromosomes, its genetic reward is greater likelihood that some of its clan will be transmitted to the next generation of hosts.

Furthermore, when TEs infest a host genome, they often do harm. Their abundant numbers probably add a metabolic burden to host cells, and TEs frequently generate harmful mutations when they happen to insert into a functional gene and disrupt its normal activities. Thus, many active TEs are not merely selfish, unruly nomads, but also parasitic DNA sequences, genuinely hurtful to their landlords. How can they then survive and spread in a host population? On balance, this happens whenever a TE gains more in its transmission success by dispersing itself across chromosomes than it loses by debilitating its host. Should we then consider TEs to be little intracellular terrorists as well? Probably not. No wise TE actually "intends" to harm its host; any damage done is simply a by-product of the element's rambunctious nature, like a bouncing kangaroo in a glass shop.

Before simply condemning these shifty intracellular agents, we should be aware of the host-serving benefits that some TEs provide. First, a few TE-induced mutations may by chance improve rather than harm a cell's functional operations. Second and more important, TEs often carry gene-regulatory sequences that over evolutionary time can occasionally be drafted into the adaptive service of a host organism, imbuing its cells with powerful new metabolic capabilities.

Jumping genes can be helpful to genetic engineers. Much like viral vectors (previous essay), some TEs can be harnessed to pick up, transport, and copy exogenous DNA into new biological sites. For example, P elements are a class of TEs in laboratory *Drosophila* that researchers have used to insert transgenic DNA sequences into these fruit flies' germlines. Another TE, known as sleeping beauty, has proved to be a suitable vector for transferring foreign genes into vertebrate animals ranging from fish to humans. In each such example, genetic engineers simply hitch a piece of DNA to a TE and let this vector do what comes naturally: jump into and then proliferate within its recipient host.

As with viruses, some risks attend the use of jumping genes as transformation vectors. Most important is the fact that genetic engineers cannot precisely direct TE movements in most cases. Thus, a TE may carry a transgene to biological destinations and may have biological consequences that the genetic engineer never intended.

Galls and Goals in Plant Transformation

A gall is a plant's version of a tumor—an unsightly swelling on stems or roots attributable to uncontrolled cell growth. Common causal agents are several species of *Agrobacterium*, rod-shaped microbes that otherwise inhabit the soil. For example, the pathogenic form of one such bacterium (*A. tumefaciens*) produces crown galls above ground, another (*A. rhizogenes*) produces root galls, and another (*A. rubi*) instigates cane galls, especially in raspberries. These bacteria primarily attack broad-leaved plants (dicots) such as grapevines, roses, and fruit and nut trees, including the apple, pear, cherry, peach, walnut, and almond. Monocot plants, which include rice, corn, and other agronomically important cereal crops, are naturally resistant, but scientists have found experimental conditions under which supervirulent forms of *A. tumefaciens* can be made to infect some of these species as well.

How and why these soil bacteria occasionally become plant invaders has long been of special interest to geneticists. The tumorigenic process starts when a plant receives a natural or induced wound. The damaged plant cells release sugars and other compounds to which the motile bacteria are attracted via chemotaxis. After a bacterium swims to the wounded site, a small portion of its DNA (T-DNA, carried on tumor-inducing *Ti* plasmids) infiltrates nearby plant cells and integrates into their chromosomes. The bacterial plasmid genes then take command of the plant's cellular operations by inciting the synthesis of hormonelike compounds (including amino acid derivatives known as opines) that richly nourish the bacterial colony but, by fostering hormonal imbalances in the plant, also promote gall formation.

These and many further details of crown gall disease had been worked out by 1982, after more than three decades of complicated detective work. The findings were revolutionary and will be listed forever among the crowning achievements of plant molecular genetics for two reasons. First, they overturned a prevailing myth that gene transfer between kingdoms was impossible. In this case, the bacteria have molecular mechanisms that routinely inject microbial genes into plants. Second, the basic scientific discoveries about crown gall syndrome carried huge ramifications for applied biotechnology. The same infectious properties that make *A. tumefaciens* an expert micro-commando for interjecting its own DNA into plant cells make it a powerful delivery system (vector) for inserting other genes too. In 1983, scientists first harnessed the services of *Agrobacterium* and its *Ti* plasmids to create transgenic plant tissues in the laboratory. Another significant molecular achievement occurred in 2001 when researchers published the complete sequence of the *Agrobacterium* genome, a milestone event that will permit a deeper understanding of this microbe's physiology.

Geneticists today continue to be preoccupied with these gall-inducing bacteria. In a typical experiment, a researcher uses molecular methods to cut out the tumor-forming part of a *Ti* plasmid (thereby disarming it) and then replaces it with a desired gene. The latter might be, for example, a microbial gene that kills particular insect pests or a gene from another plant species that conveys tolerance to colder climates. After this gene (and any appropriate regulatory escorts) has been engineered into *Ti* plasmids and returned to *A. tumefaciens*, the bacteria are simply left to do what comes naturally (i.e., infest the wounds of targeted plants and inject the exotic genes). Thereby a new variety of transgenic crop can be created that may display a novel or improved trait of interest. This is precisely how crops with resistance to insects and herbicides (see chapter 3) were engineered, and it has become the standard approach for producing transgenic plants.

The gall bacterium has largely supplanted several other plant genetic transformation techniques that were introduced at about the same time. Two of these, particle bombardment and electroporation, nonetheless deserve mention. Particle bombardment, also known as biolistics, uses a gene gun (literally a firearm in the first experiments in 1987) to shoot microscopic metal beads, coated with DNA, at target cells. Remarkably, if the muzzle velocity is adjusted properly such that the cell is not disrupted, exotic DNA can be delivered to plants in this way. Electroporation achieves the same goal by applying electrical pulses to cells. The pulses open microscopic pores in the cell membrane through which DNA molecules can enter. Although these and other genetic transformation methods still are used occasionally (especially with animal and bacterial cells), gene guns and electroporation equipment have gathered dust in most laboratories after the introduction of *A. tumefaciens* as a powerful and flexible plant transformation vector.

With all this genetic engineering attention devoted to gall bacteria as DNA delivery vehicles, have scientists neglected the plight of the tumor-ridden plants? Galls certainly do considerable harm to plants by blocking the transport of water and nutrients through affected portions of stems or branches. Indeed, crown gall disease is a major worldwide cause of lowered productivity in perennial crops. Accordingly, biotechnology companies also have been keenly interested in preventing the disease.

In the late 1990s, scientists began designing experimental transgenic tomatoes, flowering mustards, and aspen trees with the intent of making these plants resistant to pathogenic strains of *A. tumefaciens*. The approach involves introducing genes whose effect is to block the hormone-altering effects of the bacteria, thereby inhibiting gall formation. The initial experiments worked. The hope is that such tumor-resistant plants someday may become a staple of agriculture and forestry.

Promoting Promoters and Constructing Constructs

Genetic engineering can be a hit-and-miss affair. In a typical gene-engineering experiment in botany, for example, a transformation vector may deliver its DNA cargo to only a few among thousands of exposed plant cells. Furthermore, even when a transgene does incorporate into a recipient cell, it often fails to express properly (i.e., produce the desired protein) in its new molecular environment.

Inserting a transgene at random into the cells of another creature is somewhat like throwing an exotic species into a native ecosystem: The foreigner might survive and function well in its novel ecological setting, but more likely it will be functionally decrepit or positively disruptive of normal system operations. Consider, for example, all that must go right when a human transgene is placed into a goat's genome, the hope being that the GM animal will produce a valuable protein in extractable quantities for medical or pharmaceutical purposes. Within a goat's cells, that human gene must be activated properly to produce messenger RNA, which in turn must be transcribed and translated into a stable protein, which in turn must be transported or stored in convenient places in the goat's body (e.g., in milk or blood) for easy retrieval. Many things can go wrong along the way, and the wonder of genetic engineering is not so much that producing a transgenic protein sometimes fails, but rather that the endeavor so often succeeds.

Typically, a key step in the process involves coaxing the transgene to activate properly in its new cellular home, so as to produce high quantities of messenger RNA. To assist in this critical mission, the inserted transgene is often provided a molecular escort known as a promoter. This is a region of DNA to which an RNA polymerase binds and initiates transcription. Many promoters act in tissue-specific fashion, and this can be critical in enabling genetic engineers to target rather precisely where and when an associated transgene is activated within a recipient's body.

Promoters consist of various types of DNA sequences that differ greatly in their capacity to stimulate RNA polymerases to produce RNA molecules from DNA templates. For example, a CaMV-35S promoter from the cauliflower mosaic virus has proved suitable for driving the expression of foreign genes in dicots (one major group of flowering plants), whereas a ubiquitin promoter from maize is suited better for monocots (species that include grasses and grains). Often, effective promoters can elevate the expression of particular transgenes by a hundred-fold or more in targeted tissues. Choosing the proper promoter can make all the difference between success and failure in a recombinant DNA venture.

Thus, what a geneticist often engineers is not merely a transgene within a vector, but rather a small ensemble of DNA sequences that also includes intended promoters, enhancers, or other regulators of transgene activities. Such a construction is called an expression vector, or a transformation construct. Its constituent DNA parts may come from a single species, or they may be assembled in a mix-and-match fashion from different biological sources and pasted together for delivery in a suitable vector. It is not uncommon for a transgenic construct to include regulatory elements from a viral or bacterial species, hitched to a functional gene from a plant or animal species, all inserted into bacterial plasmids that then act as cloning and transformation agents for producing transgenic plants or animals of yet another species.

In addition, several related procedures are available for increasing the chances that a protein-coding transgene will express properly in its adopted cellular home. For example, a biotechnologist may engineer an inducible genetic construct designed to activate only upon exposure to a specific stimulus in the environment, such as a chemical that affects how particular proteins bind to the promoter region of a transgene. In some cases, transgenic cells (e.g., of bacteria) can be grown in bulk and then subjected to an inducer that stimulates production of a desired commercial product. Another approach in transgenic research is to use high-copy-number vectors, often present in hundreds of copies per cell. More copies of a transgene usually, but not always, mean more protein product.

In short, by constructing suitable constructs, a biotechnologist can often coax an introduced transgene to produce high levels of a particular protein, sometimes wherever or whenever desired in the host's body. In a GMO, such proteins are called foreign or heterologous because they are not native or indigenous to that creature.

Reporter Genes

How do scientists know when a transgenic cell line or creature actually has been produced? One way was mentioned in the discussion of plasmids as transformation vectors (see "Copy and Duplicate"): the lab experiments are conducted in such a way that antibiotics then can be used to screen bacterial colonies for those that are, or are not, genetically transformed. In other situations, the presence of a transgene may be evidenced by its impact on the recipient's appearance or metabolism. For example, a GM fish that carries a growth-hormone gene (see chapter 6) may grow much faster and achieve a greater weight than its non-transformed compatriots. However, few transgenes have such overt visible effects on recipient organisms, and it may take months to gauge by morphological or physiological inspection whether gene incorporation has been achieved.

A more general approach to detect successful gene transfer is to use marker genes, also known as reporter genes. These are pieces of DNA physically added to an expression vector that merely serve to notify a researcher when a particular transformation has been successful. In other words, the marker gene signals to the biotechnologist when an adjoining transgene has been incorporated successfully in a recipient cell or organism. This signal often consists of a flash of visible light from the recipient cells, like a neon sign announcing "the transgene has arrived!" The net effect of a reporter gene can be as spectacular as a tobacco plant whose leaves glow in the dark or a fruit fly whose eyes emit a bright green aura.

One popular reporter gene was isolated from the common firefly (*Photinus pyralis*), where it encodes the enzyme luciferase that gives these creatures their characteristic ability to produce and flash light. Another popular marker gene, isolated from a luminescent jellyfish (*Aequorea victoria*), produces a green fluorescent protein (GFP). A whole battery of reporter genes has been isolated from the bioluminescent click beetle, *Pyrophorus plagiophthalamus*. When inserted and expressed in bacteria, these genes code for luciferase proteins that emit four different color-coded signals: orange, green, yellow, and yellow-green.

The basic procedure is to incorporate the marker gene, together with the rest of the genetic construct of interest, into a transformation vector such as a transposable element. If this composite vector is taken up by cells of a host organism and appropriately turned on in some or all of its tissues, the luminescent proteins that are produced then light up. The researcher need only dim the room lights and look for the glow of this GM creation to assess whether the genetic transfer was accomplished successfully.

Earlier kinds of reporter genes were nonluminescent and instead often involved selectable markers. For example, a herbicide-resistance marker gene could be attached to another transgene in an experimental transformation of plants. Only the specimens that actually incorporated the genetic construct would grow well in the presence of the herbicide. Such selectable markers were less than fully satisfactory, however, because they could affect plant growth and survival; and, furthermore, there was always the possibility that the selectable marker gene might escape the original intended host and directly benefit other species (including weeds, in this case). Thus, extensive research effort has been invested in ways to excise selectable marker genes from GM plants. In contrast, luminescent markers are inert genetic tags that presumably neither help nor harm their bearers and therefore do not always require removal.

Luminescent reporter genes have been used to monitor the outcomes of genetic transformation experiments in a wide variety of species ranging from viruses, bacteria, and yeast, to plants, invertebrate animals such as fruit flies and worms, and many vertebrate species including fish, frogs, mice, cows, and monkeys. Some engineered plants carry a blue variant of GFP. When these plants are exposed to ultraviolet light, they glow purple (a combination of the blue, plus a red that appears naturally when the plant's chlorophyll is exposed to UV light). Some commonly used species of transgenic fish, such as zebrafish, are especially intriguing (see chapter 6). Their bodies are translucent, so their internal organs, bathed in fluorescence, can be observed in great detail.

Test-Tube Gene Cloning

As already discussed, scientists can make perfect replicas of small pieces of DNA by inserting these into plasmids, viruses, or mobile elements and then letting these tiny biological vectors proliferate naturally in host cells. Such DNA cloning takes place *in vivo*. Several years ago, a technological breakthrough permitted scientists to clone pieces of DNA *in vitro* for the first time, without direct involvement of live organisms. This revolutionary procedure, known as the polymerase chain reaction (PCR), has been a huge boon to biotechnology by greatly simplifying and facilitating the formerly arduous process of gene cloning. The invention of PCR is a human-interest story and also merits telling for its broader lessons about the nature of scientific discovery.

In the 1960s, microbiologists found thermophilic (heat-loving) bacteria inhabiting the hot springs of Wyoming's Yellowstone National Park. These microbes had evolved exceptional metabolic machineries that enable them to thrive at hot water temperatures where other species are scalded. One of the heat-loving bacteria, *Thermus aquaticus*, was to achieve lasting fame by donating a special heat-resistant enzyme that is a critical component of the PCR.

That enzyme, *Taq* polymerase, was isolated in 1976 by Alice Chien and colleagues. *Taq*'s normal function inside *T. aquaticus* is to catalyze replication of the bacterium's DNA, a remarkable feat requiring the enzyme to operate routinely at 74°C and remain active even after exposure to near-boiling waters (95°C). *Taq*'s capacity to replicate DNA at high temperatures was the key to *in vitro* gene cloning, a seminal insight that won Kary Mullis a Nobel Prize in 1993.

Cetus, the biotech company that Mullis worked for in the 1980s when he made his discoveries, already had developed methods for synthesizing short pieces of DNA that anneal or stick to a specific location in a genome. Mullis reasoned that if he could use two such pieces of DNA (primers) that would anneal specifically to regions flanking a particular gene of interest, he might then be able to clone that gene. All he need do is use a DNA polymerase over and over again to copy the DNA region sandwiched between the two flanking primers. A few nights of experimental tinkering in the lab demonstrated the feasibility of this idea and gave birth to the now famous polymerase chain reaction.

There are three basic steps to each round of PCR: denature, anneal, and extend. First, double-stranded DNA (from any organism) is isolated and then heated to a high temperature, resulting in its denaturation (separation into single strands). Second, two short primer sequences, each about 20 base pairs in length, are annealed or hybridized (see next essay) to specific regions flanking the longer piece of DNA to be cloned. Third, these primers are extended by a DNA polymerase that fills in complementary nucleotides between the two flanking primers and thereby reconstitutes double-stranded DNA. Via these three successive steps, each original DNA molecule becomes two identical molecules. Successive rounds of PCR then expand the number of copies to 4, 8, 16, and so on in geometric progression. After 20 rounds of PCR, each original DNA molecule has been amplified to more than a million. The whole gene-cloning procedure is conveniently accomplished in a test tube.

Taq polymerase plays a key role in PCR because it remains stable and active across each required round of high-temperature DNA denaturation. Ordinary DNA polymerases isolated from other species do not perform nearly so well in PCR because they are denatured at each high-temperature cycle. Thus, ordinary DNA polymerases must be added continually to the PCR process, and they quickly gum up the works. Furthermore, at the lower temperatures where they work best, the ordinary DNA polymerases often misfire or misprime, thus compromising the entire PCR enterprise. All of these problems are avoided with *Taq*.

At least two important historical lessons follow from the discovery of *Taq* polymerase and the invention of the highly successful PCR that it enabled. First, findings from pure or basic research often contribute in unanticipated ways to applied technology. Second, some of the earth's most inconspicuous organisms can furnish highly utilitarian biological compounds. Who could have guessed that an enzyme from an obscure heat-loving bacterium from Yellowstone Park would so revolutionize molecular biology?

Like traditional *in vivo* gene cloning, the PCR permits scientists to make virtually unlimited copies of any short piece of DNA, typically from a few hundred to several thousands of basepairs long. Unlike *in vivo* gene cloning, PCR requires no living hosts for the process. PCR is now automated and conducted routinely in small machines (the size of a bread box) that sit on the bench-tops of biological laboratories throughout the world.

Hybridizing DNA Molecules

Another natural property of DNA of great practical utility to biotechnologists is the molecule's proclivity to hybridize with itself in highly precise and specific ways. Recall that native DNA is a double helix, two complementary strands held together by weak chemical links known as hydrogen bonds that involve precise pairings between each A and T nucleotide, and between each C and G. Such complementarity between strands, stemming from the fundamental rules of nucleotide pairing, provides the biological basis for various laboratory methods, known generically as DNA hybridization.

Hydrogen bonds are the weakest part of the DNA's overall physical structure, so the two intact strands can be artificially dissociated (melted) by heating duplex DNA to near-boiling temperatures or by various chemical treatments. Suppose that a test tube contains, in solution, many such sets of melted strands. If the dissociating agent (high temperature or chemicals) then is removed, the separated strands often find their proper pairing partners and spontaneously reassociate into duplex DNAs. In other words, single strands that are complementary in nucleotide sequences naturally recognize, align, and hybridize with one another to reform intact duplexes. Such hybridization does not require that DNA strands are perfectly complementary—a modest percentage of basepair mismatches is allowed. However, by adjusting the stringency of experimental conditions, geneticists can regulate and monitor the DNA-hybridization process as a function of the degree of basepair complementarity in the reassociating molecules.

The natural propensity for complementary DNA strands to hybridize has found a wealth of applications in artificial genetic manipulation, two of which already have been mentioned. First, the protruding ends of DNA cut with a given restriction enzyme are sticky precisely because other such single-stranded termini have complementary basepairs, and this feature is critically important in many of the splicing operations of recombinant DNA technology (see "Cut and Paste"). Second, the single-stranded primers used in the first step of the PCR process (see "Test-tube Gene Cloning") work because they are designed specifically to have nucleotide sequences complementary to the precise target DNA strands on which they will later anneal.

Another important class of application for DNA hybridization involves retrieving particular genes of interest from an otherwise heterogeneous collection of unrelated DNA molecules. One rationale for such efforts was presaged in the discussion of recombinant DNA libraries (see "Copy and Duplicate"). Each of the hundreds of transgenic bacterial colonies in a genomic library likely carries a different piece of foreign DNA, and the trick is to identify the precise colony or colonies containing a cloned gene of interest. To do so, geneticists incubate the bacteria under suitable DNA hybridization conditions with a radioactively labeled probe, a small and carefully designed piece of DNA. Typically synthesized in a special machine, the probe is constructed to be complementary in nucleotide sequence to a tiny portion of the sought gene, which may be hidden anywhere in the genomic library. By finding perfect DNA-sequence matches, the probe then hybridizes to ("lights up") any transgenic colony that happens to carry the desired gene.

A related use of DNA hybridization is to identify specific DNA sequences in a gel (an artificial matrix much like gelatin in texture) made of agarose or acrylamide. Electrophoresis is a routine laboratory method by which DNA molecules of varying lengths are physically separated from one another through the use of electricity. When a heterogeneous collection of DNA sequences from any source is placed into a gel and a mild electrical current is applied, small molecules migrate through the matrix faster than large ones, and after a few hours they end up at different positions in the gel. However, the DNA molecules cannot be seen by the naked eye. In 1975, scientist E.M. Southern developed an ingenious hybridization-based method for detecting where such DNA sequences have migrated in the gel.

This method again involves the use of a radioactively labeled probe. The probe locates and specifically hybridizes to any matched DNA sequences. This time, the end result is the appearance of visible bands on the gel, each specifying the exact position to which a specific piece of DNA has migrated. The approach is called Southern blotting, in honor of its inventor.

Two related laboratory procedures, dubbed Northern blotting (used to distinguish RNA molecules) and Western blotting (for identifying specific proteins), reveal that wet-lab geneticists must have a dry sense of humor; these methods do not reflect a compass heading or geographic region, but instead derive from the term Southern blotting.

Knockouts and Resuscitations

Mutations are an enigma. Although necessary over the longer term as the ultimate source of genetic variation upon which evolutionary processes depend, not all but most newly arisen mutations are harmful or neutral to their bearers in the short term. Like an occasional replacement spring or sprocket thrown haphazardly into a well-working watch, *de novo* mutations generally disrupt rather than improve cellular operations, and most are quickly eliminated by natural selection. But, just as watches without replacement parts eventually cease to function, species require a regular supply of novel genetic variation. Without new mutations, some fraction of which do provide benefits, especially when environments change, any species sooner or later would go extinct.

This leads to an evolutionary enigma, eloquently captured in Richard Dawkins's famous image of a blind watchmaker. Why has natural selection, the blind (or at least nearsighted) directive force of evolution, permitted the continuance of mutational processes whose evident costs are high and immediate, yet whose potential benefits (which myopic selection cannot foresee) are mostly diffuse and deferred? In other words, given natural selection's immediate scrutinizing power coupled with its inability to anticipate future needs, why over the eons have species not evolved cellular mechanisms that effectively block the formation of new mutations?

One possible response is that species generally have done so. Although new mutations arise by the thousands in each cell during every round of DNA replication, most of these copying mistakes are repaired immediately by proofreading enzymes that scrupulously check for errors. These enzyme systems correct DNA damages before they do harm. Only the rare mutation (normally one or fewer per million genes in germline cells per organismal generation) survives this refined molecular screening. Just as even the best security at a busy airport may not be able to intercept 100% of dangerous baggage, so too, the argument goes, even the most sophisticated DNA repair apparatus in cells is simply unable to interdict 100% of *de novo* mutations.

A second possibility, not incompatible with the first, is that natural selection at the level of the gene in effect promotes some classes of mutation, including those that inadvertently harm the individual. A remarkable discovery in recent years is that selfish mobile elements (see "Jumping Genes") are a leading source of mutations. Notwithstanding their negative immediate impact on host genomes, jumping genes often tend to be selectively favored at the gene level by virtue of their proliferative nature.

The bottom line, ironically, is that regardless of source, mutations—the fountainhead of biodiversity and evolutionary change—also continually damage their hosts, causing untold suffering, countless disabilities, and deaths. Another irony is that this negative fact of life is a blessing for a large branch of genetic research that involves identifying genes and exploring how they function. Indeed, the initial evidence for genetic influence over any particular aspect of an organism's phenotype (its biochemical profile, physiology, development, behavior, morphology, etc.) usually comes from the discovery of mutant forms of a gene that cause gross departures from the typical condition.

Gregor Mendel began to appreciate this fact in the mid-1800s in his seminal work on the hereditary basis of discrete variation (in the heights of pea plants and the colors and shapes of their seeds). Some 70 years later, in 1933, Thomas Hunt Morgan won the Nobel Prize in Medicine for furthering the concept of the gene as the basic unit of genetic transmission, as well as function. Morgan and his colleagues discovered fruit fly mutations affecting such traits as eye color, wing shape, and age of death and then used these variants to deduce the existence of particular genes and unveil their inheritance patterns. Throughout the twentieth century, geneticists likewise relied on random mutations (whether natural or induced by radiation, temperature, or chemicals) to reveal crucial links between heredity and physiology, morphology, or behavior.

Today, geneticists still rely on the disruptive aspects of new mutations to identify genes and their cellular functions, but new tricks have been added to their research repertoire. Using various recombinant DNA methods, particular genes in experimental species such as viruses, bacteria, yeast, worms, and mice now can be mutated purposefully and their effects on organisms monitored methodically. These laboratory techniques combine and extend several of the methods introduced in earlier essays.

For example, one form of site-directed mutagenesis involves excising the native version of a gene of interest and replacing it with altered substitutes. These substitutes may be naturally occurring, or they may arise artificially from PCR reactions purposefully conducted under error-prone conditions (so that new mutations arise during the amplification process). They may even be specific pieces of DNA synthetically generated in a machine known as a DNA synthesizer. In another approach, a piece of functionless (junk) DNA is cloned and then inserted into different parts of a wild-type (normal) gene, just to see what happens. Via such DNA tinkering in the laboratory, specific phenotypes of organisms are perturbed, and by studying what went wrong in cells bearing mutations, geneticists can thereby deduce normal genetic underpinnings of the traits in question.

Like some natural mutations, many of these intentional mutations deal a serious functional blow to a gene. By throwing a molecular haymaker into a gene's normal activity, such "knockout mutations" seriously stun the gene, thereby helping to expose its formerly hidden role. To confirm that role, as well as to revive the cells, geneticists then use recombinant-DNA methods to insert various gene sequences into the mutated cells, the intent being to rescue that debilitated gene's original function. By studying details of how a cell may be resuscitated from particular knockout mutations, researchers can further assess the normal roles and vulnerabilities of specific genes. The net result is greater insight into gene functions and cellular operations.

Glossary

adaptation Any feature (e.g., morphological, physiological, behavioral) that helps an organism to survive and reproduce in a particular environment.

allele Any of the possible alternative forms of a gene. A diploid individual carries two alleles at each autosomal gene, and these can either be identical in state (in which case the individual is homozygous) or different in state (heterozygous). At each autosomal gene, a population of *N* diploid individuals harbors 2*N* alleles, many of which may differ in details of nucleotide sequence.

allergen Something that causes an allergy.

allergy An unduly sensitive state involving the immune system and resulting from exposure to particular foreign substances.

amino acid Any of about 20 different organic molecules found universally in proteins and displaying carboxyl and amino groups, that when joined together form a polypeptide.

amniocentesis A clinical procedure for prenatal genetic diagnosis in which a sterile, hollow needle is inserted through the mother's abdomen and into her uterus and used to withdraw fetal cells sloughed into amniotic fluid.

angiogenesis The growth of new blood vessels.

antibiotic A chemical substance, produced by microbes, that destroys or inhibits the growth of other organisms and that is used by the medical profession principally for the treatment of infectious diseases.
antibody A protein produced by the immune system that reacts to foreign material and thereby protects the body against invasive substances.
antigen A substance that when introduced into the body is capable of eliciting an immune response.
artificial insemination Injection of sperm into a female's reproductive tract by a syringe or other such mechanical device.
artificial selection A human-mediated analogue of natural selection in which people promote the differential survival and reproduction of plants, animals, or microbes with specific desired traits.
asexual reproduction Any form of reproduction that does not involve the fusion of sex cells (gametes).
autosome A chromosome in the nucleus other than a sex chromosome; in diploid organisms, autosomes are present in homologous pairs.

bacteria Unicellular microorganisms without a true cellular nucleus.
bacteriophage A virus that infects a bacterium.
bioremediation Environmental restoration by living organisms.
biotechnology The use of living entities or their components or products for industrial or commercial applications.
blastocyst A mammalian embryo before its implantation into the uterine wall.
blastula A hollow sphere of cells resulting from early cell cleavages from a zygote.

cancer A disease characterized by uncontrolled, abnormal cellular proliferation.
carcinogenic Cancer causing.
cell A small, membrane-bound unit of life capable of self-reproduction.
chimera An individual composed of a mixture of genetically different cells tracing to different zygotes. *See also* mosaic.
chloroplast An organelle in the cytoplasm of plant cells that contains its own DNA (cpDNA); the site of photosynthesis.
chromosome A threadlike structure within a cell that carries genes.
clone A group of genetically identical cells or organisms, all descended from a single ancestral cell or organism; to produce such genetically identical cells or organisms.
cloning vector *See* transformation vector.
coevolution The interdependent evolution of two or more species engaged in ecological relationships.
convergent evolution The independent evolution of structural, functional, or other similarities between distantly related or unrelated species.
cryopreservation Storage of biological material at ultra-low temperatures.
cytoplasm The portion of a cell outside the nucleus.

defensins Small polypeptide molecules that afford a plant or animal with some degree of protection against particular invasive microbes.

diploid A usual somatic cell condition wherein two copies of each chromosome are present.
dizygotic fraternal twins Genetically nonidentical siblings that are born at the same time but stem from two separate zygotes during a pregnancy.
DNA (deoxyribonucleic acid) The genetic material of most life forms; a double-stranded molecule composed of strings of nucleotides.
DNA/DNA hybridization A class of laboratory procedures in which single-strand stretches of polynucleotides attract and bind to homologous, complementary single strands.

ecology The study of the interrelationships among living organisms and their environments.
ecosystem A community of ecologically interacting organisms and their environment.
egg A female gamete. *See also* oocyte.
electrophoresis The movement of charged proteins or nucleic acids through a supporting gel under the influence of an electric current.
electroporation The use of electrical pulses to open pores in the outer covering of a cell through which nucleic acids or other organic compounds might enter.
embryo An organism in the early stages of development (in humans, usually up to the end of the second month of pregnancy). *See also* pre-embryo.
embryonic stem cells *See* stem cells.
endangered species A species at immediate risk of extinction.
endemic Native to, and restricted to, a particular geographic area.
endogenous Produced or naturally occurring within the body.
enzyme A catalyst (normally a protein) of a specific chemical reaction.
eugenics The ideology or practice of attempting to improve *Homo sapiens* by altering its gene pool.
eukaryote Any organism in which chromosomes are housed in a membrane-bound nucleus.
evolution In simplest terms, any change across time in the genetic composition (i.e., in gene frequencies) of populations or species.
exogenous Produced or naturally occurring outside the body.
exon A coding segment of a gene. *See also* intron.
expression *See* gene expression.
expression vector *See* transformation vector.
extinction The permanent disappearance of a population or species. *See also* endangered species.

fermentation A slow decomposition and alteration of organic substances induced by microbes.
fertilization The union of two gametes to produce a zygote.
fetus An individual at intermediate stages of development in the uterus (in humans, beginning at about the third month of pregnancy).

fitness (Darwinian) The contribution of an individual, or of a particular genotype, to the next generation relative to the contributions of other individuals, or genotypes, in the population.

forensics (genetic) Of or pertaining to the diagnosis of otherwise unknown biological material based on analysis of proteins or DNA.

fungicide A pesticide applied to fungi.

gamete A mature reproductive sex cell (egg or sperm).

gene The basic unit of heredity; usually taken to imply a sequence of nucleotides specifying production of a polypeptide or other functional product (but also can be applied to segments of DNA with unknown or unspecified function).

gene expression Activation of a gene to begin the process (RNA formation) that later may eventuate in production of a protein.

gene flow The geographic movement of genes, normally among populations within a species.

gene gun A device for propelling DNA molecules into living cells.

gene pool The sum total of all hereditary material in a population or species.

gene stacking The practice of designing a genetically modified species to carry two or more transgenes simultaneously, such as for tolerance to different herbicides.

gene therapy The insertion of a functional gene into an individual's cells with the intent of correcting a hereditary disorder.

genetic Of or pertaining to the study of heredity.

genetically modified organism (GMO) Any plant, animal, or microbe whose genes have been deliberately and directly altered by humans.

genetic drift Changes in allele frequencies in a finite population by chance sampling of gametes between generations.

genetic engineering The direct and purposeful alteration of genetic material by humans.

genome The complete genetic constitution of an organism; also can refer to a particular composite piece of DNA, such as the mitochondrial genome.

genomic library A collection of recombinant DNA molecules (cloned fragments) of an individual's genome typically housed in bacterial colonies.

genotype The genetic constitution of an individual with reference to a single gene or set of genes.

germline The lineage of cells leading to an individual's gametes.

haploid The usual condition of a gametic cell in which one copy of each chromosome is present.

heavy metals Any highly dense metal, often toxic to organisms.

hematopoietic Of or pertaining to the formation of blood.

herbicide A chemical used to kill plants.

heredity The phenomenon of familial transmission of genetic material from one generation to the next.

heterozygote A diploid organism possessing two different alleles at a specified genetic locus.

homozygote A diploid organism possessing two identical alleles at a specified genetic locus.

hormone An organic compound produced in one region of an organism and transported by blood to target cells in other parts of the body where its effects are exerted.

hybridization The successful mating of individuals belonging to genetically different populations or species.

immune system A complex network of cells, cellular products, and tissues that helps protect the body from pathogens and other foreign substances.

immunocontraception Blockage of fertility by invoking of the immune response, as, for example, by eliciting antibodies against egg or sperm proteins.

inorganic Of or pertaining to chemicals other than carbon compounds (or their derivatives) made by living organisms.

insecticide A pesticide applied to insects.

introgression The movement of genes between species via hybridization.

intron A noncoding portion of a gene. Most genes in eukaryotic organisms consist of alternating intron and exon DNA sequences.

invertebrate An animal that does not possess a backbone.

in vitro Outside the living body (e.g., in a laboratory or test tube).

in vitro fertilization (IVF) Zygote formation (union of egg and sperm) outside a living body (i.e., in a test tube), often followed by introduction of the resulting embryo to a female mammal's reproductive tract.

in vivo Within a living body.

jumping gene *See* transposable element.

junk DNA A term formerly used to describe nucleic acid sequences that do not specify a functional protein or RNA product. *See also* selfish DNA.

knockout A mutation that destroys or inactivates a specific gene function.

life cycle The sequence of events for an individual, from its origin as a zygote to its death; one generation.

ligase (DNA) An enzyme that catalyzes the binding together of DNA pieces end to end.

locus (pl. loci) A gene, a location on a chromosome.

luciferase An enzyme from a firefly that catalyzes the production of light.

Lyon effect The natural inactivation of one of the two X-chromosomes in each somatic cell of a female mammal.

marker gene *See* reporter gene.

meiosis The cellular process whereby a diploid cell divides to form haploid gametes.

metabolism The sum of all physical and chemical processes by which living matter is produced and maintained and by which cellular energy is made available.

microbe A very small organism visible only under a microscope.

mitochondria (sing. mitochondrion) Organelles in the cytoplasm of animal and plant cells that contain their own DNA (mtDNA); the site of some of the metabolic pathways involved in cellular energy production.
mitosis A process of cell division that produces daughter cells with the same chromosomal constitution as the parent cell.
mobile element *See* transposable element.
molecular diagnosis The application of DNA-level (or protein-level) assays to reveal the presence or absence of a heritable disorder.
monozygotic twins Genetically identical siblings (barring mutation) that stem from a single zygote during a pregnancy.
morphology The visible structures of organisms.
mosaic An individual that carries two or more sets of genetically different cells, regardless of sources. *See also* chimera.
mutagen Any agent causing an increase in the frequency of new hereditary variants.
mutation A change in the genetic constitution of an organism.

natural selection The differential contribution by individuals of different genotypes to the population of offspring in the next generation.
nuclear transfer (NT) cloning The construction of genetically identical organisms by an artificial process that begins with transporting the nucleus of a somatic cell into an enucleated egg cell.
nucleic acid *See* DNA and RNA.
nucleotide A unit of DNA or RNA consisting of a nitrogenous base, a pentose sugar, and a phosphate group.
nucleus (pl. nuclei) A portion of a cell bounded by a membrane and containing chromosomes.

oocyte A female gamete, also known as an egg cell, or ovum.
organelle A complex, recognizable structure in the cell cytoplasm (such as a mitochondrion or chloroplast).
organic compound The carbon-based chemicals made by living or formerly living organisms.

parasite An organism that at least for some part of its life cycle is intimately associated with and harmful to a host.
parthenogenesis The development of an individual from an unfertilized egg.
pathogen An organism or microorganism that produces a disease.
pedigree A diagram displaying population ancestry (mating partners and their offspring across generations).
pesticide A chemical agent that kills animal pests.
phage *See* bacteriophage.
photosynthesis The biochemical process by which a plant uses light to produce carbohydrates from carbon dioxide and water.
physiology Metabolic functions of living organisms.
phytoremediation Environmental restoration by plants.

plasmid A small extrachromosomal genetic element found in bacteria.
pluripotent Capable of generating multiple types of cells or tissues.
pollen A male gamete of plants.
polyembryony The production of genetically identical offspring within a clutch or litter.
polymerase An enzyme that catalyzes the formation of nucleic acid molecules.
polymerase chain reaction (PCR) A laboratory procedure for the *in vitro* replication of DNA from even small starting quantities.
polymorphism With respect to particular organismal features or genotypes, the presence of two or more distinct forms in a population.
polypeptide A string of amino acids.
polyploidy A condition in which more than two sets of chromosomes are present within a cell.
population All individuals of a species normally inhabiting a defined area.
precautionary principle A careful approach to risk management when scientific knowledge is incomplete.
predator An organism that feeds by preying on other organisms.
pre-embryo An organism in the extremely early stages of development, in mammals typically at the pre-implantation stage. *See also* embryo.
primer (for PCR) A short string of nucleotides used in conjunction with an appropriate enzyme to initiate synthesis of a nucleic acid.
prokaryote Any microorganism that lacks a chromosome-containing, membrane-bound nucleus.
promoter A region of DNA to which an RNA polymerase binds and initiates transcription.
protein A macromolecule composed of one or more polypeptide chains.

recombinant DNA A new hereditary molecule that has arisen from genetic recombination.
recombination (genetic) The formation of new combinations of genes, as, for example, occurs naturally via meiosis and fertilization.
regulatory gene A segment of DNA that exerts operational control over the expression of other genes.
reporter gene A piece of DNA, added to a transformation vector, that signals when the transformation was successful.
reproductive cloning The construction of genetically identical organisms via biotechnology.
restriction enzyme Any organic compound produced by a bacterium that catalyzes the cleavage of DNA molecules at specific recognition sites.
restriction fragment A linear segment of DNA resulting from cleavage of a longer segment by a restriction enzyme.
retrotransposable element A form of jumping gene that tranposes via an RNA intermediate.
retrovirus An RNA virus that uses reverse transcription during its life cycle to integrate into the DNA of host cells.

reverse transcription DNA synthesis from an RNA template.

RNA (ribonucleic acid) The genetic material of many viruses, similar in structure to DNA. Also, any of a class of molecules that normally arise in cells from the transcription of DNA.

selectable marker A gene that by virtue of responsiveness to specific experimental growing conditions can indicate its presence to the researcher (i.e., by protecting the cell that houses it from an antibiotic or a toxic chemical). *See also* reporter gene.

selfish DNA Hereditary material that displays self-perpetuating modes of behavior without apparent benefit to the organism. *See also* junk DNA.

senescence A persistent decline with age in the survival probability or reproductive output of an individual due to physiological deterioration.

sex chromosome A chromosome in the cell nucleus involved in determination of sex.

sexual reproduction Organismal procreation via the generation and subsequent fusion of gametes.

somatic Of or pertaining to any cell (or body part) in a multicellular organism other than those destined to become gametes.

species (biological) Groups of actually or potentially interbreeding individuals that are reproductively isolated from other such groups.

sperm A male gamete in animals.

stem cells Undifferentiated, mitotically active cells that serve to produce new cells or replenish those lost during the life of an individual. Those found in mature organisms are termed adult stem cells; those found in early life stages are termed embryonic stem cells.

telomere The end or tip region of a chromosome, usually containing repetitive DNA sequences.

therapeutic cloning The construction of genetically identical cells, via biotechnology, with the intent of producing replacement cells or tissues.

tissue A population of cells of the same type performing the same function.

totipotent Pertaining to cells capable of generating an entire organism.

toxin A poisonous substance.

transcription The cellular process by which an RNA molecule is formed from a DNA template.

transformation The introduction of foreign DNA into a cell or organism.

transformation vector An ensemble of DNA sequences that includes at least one transgene, one or more regulators of gene expression, and perhaps other elements such as reporter genes.

transgene Foreign DNA carried by a genetically modified organism.

transgenic organism A genetically engineered organism containing foreign DNA.

translation The cellular process by which a polypeptide chain is formed from an RNA template.

transposable element Any of a class of DNA sequences that can move from one chromosomal site to another, often replicatively.

transpose The act of transposition.

transposition (genetic) The movement of a piece of DNA from one chromosomal location to another.

transposon *See* transposable element.

triploidy A polyploid condition in which three sets of chromosomes are present in a cell.

uterus The mammalian womb.

vaccine Ideally, a harmless biological agent, prepared from a pathogen, that when delivered to an animal elicits an immune response providing the individual with protection against the disease.

vector *See* transformation vector.

vertebrate An animal that possesses a backbone.

virus A tiny, obligate intracellular parasite, incapable of autonomous replication, that uses the host cell's replicative machinery.

vitamin An organic dietary compound required in relatively minute amounts for normal growth and health.

X chromosome The sex chromosome normally present as two copies in female mammals (the homogametic sex), but as only one copy in males (the heterogametic sex).

xenobiotic A chemical or substance that is foreign to an organism.

xenotransplantation The surgical removal of an organ or tissue from one species and its transfer to a member of a different species.

Y chromosome In mammals, the sex chromosome normally present in males only.

yield drag A common phenomenon of reduced productivity in transgenic crops.

zygote Fertilized egg; the diploid cell arising from the union of male and female haploid gametes.

References and Further Readings

Preface

van Kolfschooten, F. 2002. Can you believe what you read? *Nature* 416:360–363.

1 A Tale of Good and a Tale of Evil

A New Papaya

Fitch, M.M., R.M. Manshardt, D. Gonsalves, J.L. Slightom, and J.C. Sanford. 1992. Virus resistant papaya derived from tissues bombarded with the coat protein gene of papaya ringspot virus. *Bio/Technology* 10:1466–1472.

Gonsalves, D. 1998. Control of papaya ringspot virus in papaya: a case study. *Annu. Rev. Phytopathol.* 70:1028–1032.

Microbiological Terrorism

Alibek, K. 1999. *Biohazard: The Chilling True Story of the Largest Covert Biological Weapons Program in the World.* Random House, New York.

Diamond, J. 1999. *Guns, Germs, and Steel.* W.W. Norton, New York.

Enserink, M. 2002. On biowarfare's frontline. *Science* 296:1954–1956.

Ewald, P. 2000. *Plague Time: How Stealth Infections Are Causing Cancers, Heart Disease, and Other Deadly Ailments*. Free Press, New York.
Fenn, E.A. 2001. *Pox Americana: The Great Smallpox Epidemic of 1775–82*. Hill and Wang, New York.
Finkel, E. 2001. Engineered mouse virus spurs bioweapons fears. *Science* 291:585.
Fraser, C.M., and M.R. Dando. 2001. Genomics and future biological weapons: the need for preventive action by the biomedical community. *Nature Genet.* 29:253–256.
Knobler, S.L., A.A.F. Mahmoud, and L.A. Pray, editors. 2002. *Biological Threats and Terrorism*. National Academy Press, Washington, DC.
Maynard, J.A. and others. 2002. Protection against anthrax toxin by recombinant antibody fragments correlates with antigen affinity. *Nature Biotech.* 20:597–601.
Orent, W. 2001. Will the black death return? *Discover* 22(11):72–77.
Parkhill, J. and others. 2001. Genome sequence of *Yersinia pestis*, the causative agent of plague. *Nature* 413:523–527.
Pickrell, J. 2001. Imperial College fined over hybrid virus risk. *Science* 293:779–780.

2 Framework of an Unfolding Revolution

Avise, J.C. 1998. *The Genetic Gods: Evolution and Belief in Human Affairs*. Harvard University Press, Cambridge, MA.
Nicholl, D.S.T. 2002. *An Introduction to Genetic Engineering*, 2nd ed. Cambridge University Press, Cambridge, UK.
Tokar, B., editor. 2001. *Redesigning Life? The Worldwide Challenge to Genetic Engineering*. Zed Books, New York.

3 Engineering Microbes

Insulin Factories

Bliss, M. 1982 *The Discovery of Insulin*. University of Chicago Press, Chicago.
Goeddel, D.V. and others. 1979. Expression in *Escherichia coli* of chemically synthesized genes for human insulin. *Proc. Natl. Acad. Sci. USA* 76:106–110.
Hall, S.S. 1987. *Invisible Frontiers*. The Atlantic Monthly Press, New York.
Marshall, E. 1997. A bitter battle over insulin gene. *Science* 277:1028–1030.
Ullrich, A. and others. 1977. Rat insulin genes: construction of plasmids containing coding sequences. *Science* 196:1313–1319.
Villa-Komaroff, L. and others. 1978. A bacterial clone synthesizing proinsulin. *Proc. Natl. Acad. Sci. USA* 75:3727–3731.

A Growth Industry

Barinaga, M. 1999. Genentech, UC settle suit for $200 million. *Science* 286:1655.
Goeddel, D.V. and others. 1979. Direct expression in *Escherichia coli* of a DNA sequence coding for human growth hormone. *Nature* 281:544–548.
Raben, M.S. 1958. Treatment of a pituitary dwarf with human growth hormone. *J. Clin. Endocrinol. Metab.* 18:901–903.
Tatterall, R. 1996. A history of growth hormone. *Hormone Res.* 46:236–247.

Microbial Factories for Pharmaceutical Drugs

Derynck, R. and others. 1980. Expression of a human fibroblast interferon gene in *Escherichia coli*. *Nature* 287:193–197.

Goeddel, D.V. and others. 1980. Human leukocyte interferon produced by *E. coli* is biologically active. *Nature* 287:411–415.

Itakura, K. and others. 1977. Expression in *Escherichia coli* of a chemically synthesized gene for the hormone somatostatin. *Science* 198:1056–1063.

Rai, A.K., and R.S. Eiserberg. 2003. Bayh-Dole reform and the progress of biomedicine. *Am. Sci.* 91:52–59.

More Industrious Microbes

Levine, H. III. 1999. *Genetic Engineering: A Reference Handbook*. ABC-CLIO, Santa Barbara, CA.

McGloughlin, M.N., and J.I. Burke. 2000. *Biotechnology: Present Position and Future Development*. Teagasc Publishers, Dublin, Ireland.

Accelerated, Directed Evolution

Crameri, A., S.-A. Raillard, E. Bermudez, and W.P.C. Stemmer. 1998. DNA shuffling of a family of genes from diverse species accelerates directed evolution. *Nature* 391:288–291.

Gilbert, W. 1978. Why genes in pieces? *Nature* 271:501.

Kolkman, J.A., and W.P.C. Stemmer. 2001. Directed evolution of proteins by exon shuffling. *Nature Biotech.* 19:423–428.

Patten, P.A., R.J. Howard, and W.P.C. Stemmer. 1997. Applications of DNA shuffling to pharmaceuticals and vaccines. *Curr. Opin. Biotech.* 8:724–733.

Punnonen, J., R.G. Whalen, P.A. Patten, and W.P.C. Stemmer. 2000. Molecular breeding by DNA shuffling. *Sci. and Med.* 7(2):38–47.

Stemmer, W.P.C. 1994. Rapid evolution of a protein *in vitro* by DNA shuffling. *Nature* 370:389–391.

Tobin, M.B., C. Gustafsson, and G.W. Huisman. 2000. Directed evolution: the “rational” basis for “irrational” design. *Curr. Opin. Struct. Biol.* 10:421–427.

Zhang, Y.-X., K. Perry, V.A. Vinci, K. Powell, W.P.C. Stemmer, and S.B. del Cardayré. 2002. Genome shuffling leads to rapid phenotypic improvement in bacteria. *Nature* 415:644–646.

4 Getting Creative with Crops

Gaskell, G., M.W. Bauer, J. Durant, and N.C. Allum. 1999. Worlds apart? The reception of genetically modified foods in Europe and the U.S. *Science* 285:384–387.

Hails, R.S. 2000. Genetically modified plants—the debate continues. *Trends Ecol. Evol.* 15:14–18.

Huang, J., S. Rozelle, C. Pray, and Q. Wang. 2002. Plant bioengineering in China. *Science* 295:674–677.

Ledoux, L., and R. Huart. 1968. Integration and replication of DNA of *M. lysodeikticus* in DNA of germinating barley. *Nature* 218:1256–1259.

McHughen, A. 2000. *Pandora's Picnic Basket: The Potential and Hazards of Genetically Modified Foods*. Oxford University Press, New York.

National Academy of Sciences. 1987. *Introduction of Recombinant DNA-Engineered Organisms into the Environment: Key Issues*. National Academy Press, Washington, DC.

National Academy of Sciences. 2000. *Genetically Modified Pest-Protected Plants: Science and Regulation.* National Academy Press, Washington, DC.

Nottingham, S. 1998. *Eat Your Genes: How Genetically Modified Food is Entering Our Diet.* Zed Books, London.

Office of Science and Technology. 1986. Coordinated Framework for the Regulation of Biotechnology. U.S. Government Printing Office, Washington, DC.

Rissler, J., and M. Mellon. 1996. *The Ecological Risks of Engineered Crops*. The MIT Press, Cambridge, MA.

Wolfenbarger, L.L., and P.R. Phifer. 2000. The ecological risks and benefits of genetically engineered plants. *Science* 290:2088–2093.

Combating Corn Borers

Brown, K. 2001. Seeds of concern. *Sci. Am.* 284(4):52–57.

Carson, R. 1962. *Silent Spring*. Houghton-Mifflin, Boston.

Gatehouse, A.M.R, N. Ferry, and R.J.M. Raemaekers. 2002. The case of the monarch butterfly: a verdict is returned. *Trends Genet.* 18:249–251.

Losey, J.E., L.S. Rayor, and M.E. Carter. 1999. Transgenic pollen harms monarch larvae. *Nature* 399:214.

Obrycki, J.J., J.E. Losey, O.R. Taylor, and L.C.H. Jesse. 2001. Transgenic insecticidal corn: beyond insecticidal toxicity to ecological complexity. *BioScience* 51:353–361.

Palumbi, S.R. 2001. Humans as the world's greatest evolutionary force. *Science* 293:1786–1790.

Pimentel, D.S., and P.H. Raven. 2000. *Bt* corn pollen impacts on nontarget Lepidoptera: assessment of effects in nature. *Proc. Natl. Acad. Sci. USA* 97:8198–8199.

Sears, M.K. and others. 2001. Impact of *Bt* corn pollen on monarch butterfly populations: a risk assessment. *Proc. Natl. Acad. Sci. USA* 98:11937–11942.

Wraight, C.L., A.R. Zanger, M.J. Carroll, and M.R. Berenbaum. 2000. Absence of toxicity of *Bacillus thuringiensis* pollen to black swallowtails under field conditions. *Proc. Natl. Acad. Sci. USA* 97:7700–7703.

Insecticidal Cotton

Benbrook, C.M., E. Groth, J.M. Hoaaloran, M.K. Hansen, and S. Marquardt. 1996. *Pest Management at the Crossroads*. Consumers Union, Yonkers, NY.

Gahan, L.J., F. Gould, and D.G. Heckel. 2001. Identification of a gene associated with *Bt* resistance in *Heliothus virescens*. *Science* 293:857–860.

Griffitts, J.S., J.L. Whitacre, D.E. Stevens, and R.V. Aroian. 2001. Bt toxin resistance from loss of a putative carbohydrate-modifying enzyme. *Science* 293:860–864.

Liu, Y.-B., B.E. Tabashnik, T.J. Dennehy, A.L. Patin, and A.C. Bartlett. 1999. Developmental time and resistance to Bt crops. *Nature* 400:519.

National Research Council. 2000. *Genetically Modified Pest-protected Plants: Science and Regulation.* National Academy Press, Washington, DC.

Pimental, D., and H. Lehman, editors. 1993. *The Pesticide Question: Environment, Economics and Ethics.* Chapman & Hall, New York.

Rausher, M.D. 2001. Co-evolution and plant resistance to natural enemies. *Nature* 411:857–864.

Shelton, A.M., J.D. Tang, R.T. Roush, T.D. Metz, and E.D. Earle. 2000. Field tests on managing resistance to *Bt*-engineered plants. *Nature Biotech.* 18:339–342.

Tabashnik, B.E., Y.-B. Liu, N. Finson, L. Masson, and D.G. Heckel. 1997. One gene in diamondback moth confers resistance to four *Bacillus thuringiensis* toxins. *Proc. Natl. Acad. Sci. USA* 94:1640–1644.

Defensins and Potato Famines

Almeida, M.S., K.M. Cabral, R.B. Zingali, and E. Kurtenback. 2000. Characterization of two novel defense proteins from pea (*Pisum sativum*) seeds. *Arch. Biochem. Biophys.* 378:278–286.

Andreu, D., and L. Rivas. 1998. Animal antimicrobial peptides: an overview. *Biopolymers* 47:415–433.

Coghlan, A. 1999. Fighting blight. *New Scientist* 164(2216):8.

Gao, A.G. and others. 2000. Fungal pathogen protection in potato by expression of a plant defensin peptide. *Nature Biotech.* 18:1307–1310.

Garciá-Olmedo, F., A. Molina, J.M. Alamillo, and P. Rodríguez-Palenzuéla. 1998. Plant defense peptides. *Biopolymers* 47:479–491.

Hughes, A.L. 1999. Evolutionary diversification of the mammalian defensins. *Cell. Mol. Life Sci.* 56:94–103.

Lamberty, M. and others. 1999. Insect immunity: isolation from the lepidopteran *Heliothis virescens* of a novel insect defensin with potent antifungal activity. *J. Biol. Chem.* 274:9320–9326.

Osusky, M., G. Zhou, L. Osuska, R.E. Hancock, W.W. Kay, and S. Misra. 2000. Transgenic plants expressing cationic peptide chimeras exhibit broad-spectrum resistance to phytopathogens. *Nature Biotech.* 18:1162–1166.

Zasloff, M. 2002. Antimicrobial peptides of multicellular organisms. *Nature* 415:389–395.

Herbicide-tolerant Soybeans

Dekker, J., and S.O. Duke. 1995. Herbicide-resistant field crops. *Adv. Agron.* 54:69–116.

Lewis, W.J., J.C. van Lenteren, S.C. Phatak, and J.H. Tumlinson, III. 1997. A total system approach to sustainable pest management. *Proc. Natl. Acad. Sci. USA* 94:12243–12248.

Liebman, M., and E.C. Brummer. 2000. *Impacts of Herbicide Resistant Crops.* Discussion paper presented at an International Workshop on Ecological Impacts of Transgenic Crops, March 2–4, 2000, Berkeley, CA.

Naeem, S., and S. Li. 1997. Biodiversity enhances ecosystem reliability. *Nature* 390:507–509.

Thies, C., and T. Tscharntke. 1999. Landscape structure and biological control in agroecosystems. *Science* 285:893–895.

Tilman, D., D. Wedin, and J. Knops. 1996. Productivity and sustainability influences by biodiversity in grassland ecosystems. *Nature* 379:718–720.

Herbicide-resistant Weeds

Burnet, M.W.M., Q. Hart, J.A.M. Holtum, and S.B. Powles. 1994. Resistance to nine herbicide classes in a population of rigid ryegrass (*Lolium rigidum*). *Weed Sci.* 42:369–377.

Ferber, D. 1999. GM crops in the cross hairs. *Science* 286:1662–1666.

Hayes, T.B. and others. 2002. Hermaphroditic, demasculinized frogs after exposure to the herbicide atrazine at low ecologically relevant doses. *Proc. Natl. Acad. Sci. USA* 99:5476–5480.

Holt, J.S. 1992. History and identification of herbicide-resistant weeds. *Weed Technol.* 6:615–620.

MacKenzie, D. 2000. Stray genes highlight superweed danger. *New Scientist* 168(2261):6.

Snow, A.A. 2002. Transgenic crops—why gene flow matters. *Nature Biotech.* 20:542.

Traynor, P.L., and J.H. Westwood, editors. 1999. *Ecological Effects of Pest Resistance Genes in Managed Ecosystems.* Virginia Polytechnic Institute Information Systems for Biotechnology, Blacksburg, VA.

Terminator Technology

Conway, G. 2000. Genetically modified crops: risks and promise. *Conserv. Ecol.* 4(1):5–16.

Kaiser, J. 2000. USDA to commercialize "terminator" technology. *Science* 289: 709–710.

Kuvshinov, V., K. Koivu, A. Kanerva, and E. Pehu. 2001. Molecular control of transgene escape from genetically modified plants. *Plant Science* 160:517–522.

Lambrecht, B. 2001. *Dinner at the New Gene Café.* St. Martin's Press, New York.

Service, R.F. 1998. Seed-sterilizing "Terminator technology" sows discord. *Science* 282:850–851.

Chloroplast Concoctions

Daniell, H. 1999. GM crops: public perception and scientific solutions. *Trends Plant Sci.* 4:467–469.

Daniell, H. 1999. New tools for chloroplast genetic engineering. *Nature Biotech.* 17:855–856.

Daniell, H. 2002. Molecular strategies for gene containment in transgenic crops. *Nature Biotech.* 20:581–586.

Duke, S.O., editor. 1996. *Herbicide Resistant Crops: Agricultural, Economic, Environmental, Regulatory and Technological Aspects.* CRC Press, Boca Raton, FL.

Hudson, L.C., D. Chamberlain, and C.N. Stewart, Jr. 2001. GFP-tagged pollen to monitor pollen flow of transgenic plants. *Mol. Ecol. Notes* 1:321–324.
Metz, M., and J. Fütterer. 2002. Suspect evidence of transgenic contamination. *Nature* 416:600–601.
Quist, D., and I.H. Chapela. 2001. Transgenic DNA introgressed into traditional maize landraces in Oaxaca, Mexico. *Nature* 414:541–543.
Staub, J.M. and others. 2000. High-yield production of a human therapeutic protein in tobacco chloroplasts. *Nature Biotech.* 18:333–338.
van Bel, A., J. Hibberd, D. Prüfer, and M. Knoblauch. 2001. Novel approach in plastid transformation. *Curr. Opin. Biotech.* 12:144–149.

Alleviating Dietary Deficiencies: The Golden Rice Story

Grusak, M.A., and D. DellaPenna. 1999. Improving the nutrient composition of plants to enhance human nutrition and health. *Annu. Rev. Plant Physiol. Plant Mol. Biol.* 50:133–161.
World Health Organization. 1999. *Nutrition for Health and Development: Progress and Prospects on the Eve of the 21st Century*. World Health Organization, Geneva.
Ye, X. and others. 2000. Engineering the provitamin A (β-carotene) biosynthetic pathway into (carotenoid-free) rice endosperm. *Science* 287:303–305.

Supplementing Dietary Supplements: Nutrient Boosts

Feskanich, D., V. Singh, W.C. Willett, and G.A. Colditz. 2002. Vitamin A intake and hip fractures among postmenopausal women. *JAMA* 287:47–54.
Goto, F., T. Yoshihara, N. Shigemoto, S. Toki, and F. Takaiwa. 1999. Iron fortification of rice seed by the soybean ferritin gene. *Nature Biotech.* 17:282–286.
National Research Council, Food and Nutrition Board. 1989. *Recommended Daily Allowances*. National Academy Press, Washington, DC.
Shintani, D., and D. DellaPenna. 1998. Elevating the vitamin E content of plants through metabolic engineering. *Science* 282:2098–2100.
Traber, M.G., and H. Sies. 1996. Vitamin E in humans: demand and delivery *Annu. Rev. Nutrition* 16:321–347.

Going Bananas with Vaccines

Arntzen, C.J. 1995. Oral immunization with a recombinant bacterial antigen produced in transgenic plants. *Science* 268:714–716.
Carrillo, C. and others. 1998. Protective immune response to foot-and-mouth disease virus with VP1 expressed in transgenic plants. *J. Virol.* 72:1688–1690.
Carter, J.E., and W.H.R. Langridge. 2002. Plant-based vaccines for protection against infectious and autoimmune diseases. *Crit. Rev. Plant Sci.* 21:93–109.
Kapusta, J. 1999. A plant-derived edible vaccine against hepatitis B virus. *FASEB J.* 13:1796–1799.
Katz, S.L. 1997. Future vaccines and a global perspective. *The Lancet* 350:1767–1770.

Langridge, W.H.R. 2000. Edible vaccines. *Sci. Am.* 283(3):66–71.
Streatfield, S.J. and others. 2001. Plant-based vaccines: unique advantages. *Vaccine* 19:2742–2748.
Tacket, C.O., H.S. Mason, G. Losonsky, M.K. Estes, M.M. Levine, and C.J. Arntzen. 2000. Human immune responses to a novel Norwalk virus vaccine delivered in transgenic potatoes. *J. Infect. Dis.* 182:302–305.
Walmsley, A.M., and C.J. Arntzen. 2000. Plants for delivery of edible vaccines. *Curr. Opin. Biotech.* 11:126–129.

Plantibiotics and Pharmaceutical Farming

Daniell, H., S.J. Streatfield, and K. Wycoff. 2001. Medical molecular farming: production of antibodies, biopharmaceuticals and edible vaccines in plants. *Trends Plant Sci.* 6:219–226.
Dieryck, W. and others. 1997. Human haemoglobin from transgenic tobacco. *Nature* 386:29–30.
Fischer, R., S. Schillberg, and N. Emans. 2001. Molecular farming of medicines: a field of growing promise. *Outlook on Agriculture* 30:31–36.
Gruber, V. and others. 2001. Large-scale production of a therapeutic protein in transgenic tobacco plants: effect of subcellular targeting on quality of a recombinant dog gastric lipase. *Mol. Breeding* 7:329–340.
Kusnadi, A. and others. 1997. Production of recombinant proteins in transgenic plants: practical considerations. *Biotechnol. Bioeng.* 56:473–484.
Ma, J.K. and others. 1995. Generation and assembly of secretory antibodies in plants. *Science* 268:716–719.

Alleviating Allergies

Bhalla, P.L., I. Swoboda, and M.B. Singh. 1999. Antisense-mediated silencing of a gene encoding a major ryegrass pollen allergen. *Proc. Natl. Acad. Sci. USA* 96:11676–11680.
Bindslev-Jensen, C. 1998, Allergy risks of genetically engineered foods, *Allergy* 53:58–61.
Buchanan, B.B. and others. 1997. Thioredoxin-linked mitigation of allergic responses to wheat. *Proc. Natl. Acad. Sci. USA* 94:5372–5377.
Hopkin, K. 2001. The risks on the table. *Sci. Amer.* 284(4):60–61.
Nakamura, R., and T. Matsuda. 1996. Rice allergenic protein and molecular-genetic approach for hypoallergenic rice. *Biosci. Biotech. Biochem.* 60:1215–1221.
Sampson, H.A. 1997. Food allergy. *JAMA* 278:1888–1894.

Plastics from Plants

Gerngross, T.U., and S.C. Slater. 2000. How green are green plastics? *Sci. Am.* 238(2):36–41.
John, M.E., and G. Keller. 1996. Metabolic pathway engineering in cotton: biosynthesis of polyhydroxybutyrate in fiber cells. *Proc. Natl. Acad. Sci. USA* 93:12768–12773.

Kroschwitz, J.I., editor. 1990. *Polymers: Fibers and Textiles, A Compendium.* Wiley, New York.

Poirier, Y., D.E. Dennis, K. Klomparens, and C. Somerville. 1992. Polyhydroxybutyrate, a biodegradable thermoplastic, produced in transgenic plants. *Science* 256:520–523.

The Flavr Savr Tomato

Jorgensen, R.A., R.G. Atkinson, R.L.S. Forster, and W.J. Lucas. 1998. An RNA-based information superhighway in plants. *Science* 279:1486–1487.

Martineau, B. 2001. Food fight. *Sciences* (New York) 41(2):24–29.

Redenbaugh, K., W. Hiatt, B. Martineau, and D. Emlay. 1994. Determination of the safety of genetically engineered crops. *Genetically Modified Foods ACS Symposium Ser.* 605:72–87.

Sheehy, R.E., M. Kramer, and W.R. Hiatt. 1988. Reduction of polygalacturonase activity in tomato fruit by antisense RNA. *Proc. Natl. Acad. Sci. USA* 85:8805–8809.

Wolffe, A.P., and M.A. Matzke. 1999. Epigenetics: regulation through repression. *Science* 286:481–486.

A Cornucopia of GM Products

Mehta, R.A., T. Cassol, N. Li, N. Ali, A.K. Handa, and A.K. Mattoo. 2002. Engineered polyamine accumulation in tomato enhances phytonutrient content, juice quality, and vine life. *Nature Biotech.* 20:613–618.

Pew Foundation. 2001. *Harvest on the Horizon: Future Uses of Biotechnology*. Pew Charitable Trusts, Washington, DC.

Concluding Thoughts

Abelson, P.H., and P.J. Hines, editors. 1999. The plant revolution. *Science* 285:367–389.

Charles, D. 2001. *Lords of the Harvest: Biotech, Big Money, and the Future of Food.* Perseus, Cambridge, MA.

Dale, P.J., B. Clarke, and E.M.G. Fontes. 2002. Potential for the environmental impact of transgenic crops. *Nature Biotech.* 20:567–574.

Fernandez-Cornejo, J., and W.D. McBride. 2002. *Adoption of Bioengineered Crops.* U.S. Department of Agriculture, Agricultural Economic Report no. 810. USDA, Washington, DC.

Gianessi, L.P., C.S. Silvers, S. Sankula, and J.E. Carpenter. 2002. *Plant Biotechnology: Current and Potential Impact for Improving Pest Management in U.S. Agriculture*. NCFAP, Washington, DC.

Jordan, C.F. 2002. Genetic engineering, the farm crisis, and world hunger. *BioScience* 52:523–529.

Qaim, M., and D. Zilberman. 2003. Yield effects of genetically modified crops in developing countries. *Science* 299:900–902.

Smyth, S., G.G. Khachatourians, and P.W.B. Phillips. 2002. Liabilities and economics of transgenic crops. *Nature Biotech.* 20:537–541.

Thompson, J.A. 2002. *Genes for Africa: Genetically Modified Crops in the Developing World.* University of Cape Town Press, Cape Town, South Africa.

5 Genetic Engineering in the Barnyard

Avise, J.C. 2002. *Genetics in the Wild.* Smithsonian Institution Press, Washington, DC.
Dawley, R.M., and J.P. Bogart, editors. 1989. *Evolution and Ecology of Unisexual Vertebrates.* New York State Museum, Albany, NY.
Hammer, R.E. and others. 1985. Production of transgenic rabbits, sheep and pigs by microinjection. *Nature* 315:680–683.
Houdebine, L.M., editor. 1997. *Transgenic Animals: Generation and Use.* Harwood, Amsterdam.
Letourneau, D.K., and B.E. Burrows, editors. 2002. *Genetically Engineered Organisms.* CRC Press, Boca Raton, FL.
Loughry, W.J., P.A. Prödolh, C.M. McDonough, and J.C. Avise. 1998. Polyembryony in armadillos. *Am. Sci.* 86:274–279.
Murray, J.D., G.B. Anderson, A.M. Oberbauer, and M.M. McGloughlin, editors. 1999. *Transgenic Animals in Agriculture.* CABI Publishers, New York.

Spider's Silk from Goat's Milk

Guerette, P.A., D.G. Ginzinger, B.H.F. Weber, and J.M. Gosline. 1996. Silk properties determined by gland-specific expression of a spider fibrion gene family. *Science* 272:112–115.
Hillyard, P.D. 1994. *The Book of the Spider.* Random House, New York.
Hinman, M.B., J.A. Jones, and R.V. Lewis. 2000. Synthetic spider silk: a modular fiber. *TIBTECH* 18:374–379.
Kaplan, D.L. 2002. Spiderless spider webs. *Nature Biotech.* 20:239–240.
Kunzig, R. 2001. Arachnomania. *Discover* 22(9):26–27.
Lazaris, A. and others. 2002. Spider silk fibers spun from soluble recombinant silk produced in mammalian cells. *Science* 295:472–476.
Scheller, J., K.-H. Gührs, F. Grosse, and U. Conrad. 2001. Production of spider silk proteins in tobacco and potato. *Nature Biotech.* 19:573–577.
Service, R.F. 2002. Mammalian cells spin a spidery new yarn. *Science* 295:419–421.

Low-Phosphorus Enviropigs

Abelson, P.H. 1999. A potential phosphate crisis. *Science* 283:2015.
Brinch-Pedersen, H., L.D. Sørensen, and P.B. Holm. 2002. Engineering crop plants: getting a handle on phosphate. *Trends Plant Sci.* 7(3):118–125.
Golovan, S.P. and others. 2001. Pigs expressing salivary phytase produce low-phosphorus manure. *Nature Biotech.* 19:741–745.
Jongbloed, A.W., and N.P. Lenis. 1998. Environmental concerns about animal manure. *J. Anim. Sci.* 76:2641–2648.
Lei, X.G., and C.H. Stahl. 2001. Biotechnological development of effective phytases for mineral nutrition and environmental protection. *Appl. Microbiol. Biotechnol.* 57:474–481.

Mallin, M.A. 2000. Impacts of industrial animal production on rivers and estuaries. *Am. Scientist* 88(1):26–37.

Mice as Basic Research Models

Bishop, J. 1999. *Transgenic Mammals*. Pearson Education Ltd., Essex, England.

Bradley, A., M. Evans, M.H. Kaufman, and E. Robertson. 1984. Formation of germ-line chimaeras from embryo-derived teratocarcinoma cell lines. *Nature* 309:255–256.

Evans, M.J., and M.H. Kaufman. 1981. Establishment in culture of pluripotent cells from mouse embryos. *Nature* 292:154–156.

Gordon, J.W., G.A. Scangos, D.J. Plotkin, J.A. Barbosa, and F.H. Ruddle. 1980. Genetic transformation of mouse embryos by microinjection of purified DNA. *Proc. Natl. Acad. Sci. USA* 77:7380–7384.

Illmensee, K., and P.C. Hoppe. 1981. Nuclear transplantation in *Mus musculus*: developmental potential of nuclei from preimplantation embryos. *Cell* 23:9–18.

Palmiter, R.D. and others. 1982. Dramatic growth of mice that develop from eggs microinjected with metallothionein-growth hormone fusion genes. *Nature* 300:611–615.

Wakayama, T., A.C.F. Perry, M. Zuccotti, K.R. Johnson, and R. Yanagimachi. 1998. Full-term development of mice from enucleated oocytes injected with cumulus cell nuclei. *Nature* 394:369–374.

Hello Dolly

Colman, A. 2000. Somatic cell nuclear transfer in mammals: progress and applications. *Cloning* 1:185–200.

Gurdon, J.B., R.A. Laskey, and O.R. Reeves. 1975. The developmental capacity of nuclei transplanted from keratinized skin cells of adult frogs. *J. Embryol. Exp. Morph.* 34:93–112.

Kolata, G. 1998. *Clone: The Road to Dolly and the Path Ahead.* Morrow and Company, New York.

Westhusin, M.E. and others. 2001. Cloning to reproduce desired genotypes. *Theriogenology* 55:35–49.

Wilmut, I., A.E. Schnieke, J. McWhir, A.J. Kind, and K.H.S. Campbell. 1997. Viable offspring derived from fetal and adult mammalian cells. *Nature* 385:810–813.

Cow Clones

Cibelli, J.B. and others. 1998. Cloned transgenic calves produced from nonquiescent fetal fibroblasts. *Science* 280:1256–1258.

Kato, Y. and others. 1998. Eight calves cloned from somatic cells of a single adult. *Science* 282:2095–2098.

Kubota, C. and others. 2000. Six cloned calves produced from adult fibroblast cells after long-term culture. *Proc. Natl. Acad. Sci. USA* 97:990–995.

Lanza, R.P. and others. 2001. Cloned cattle can be healthy and normal. *Science* 294:1893–1894.

Westhusen, M.E. and others. 2001. Cloning to reproduce desired genotypes. *Theriogenology* 55:35–49.

Barnyard Bioreactors

Boyce, N. 2000. Lifeline in peril. *New Scientist* 168(2260):14.

Renaville, R., and A. Burny, editors. 2001. *Biotechnology in Animal Husbandry*. Kluwer, Dordrecht, The Netherlands.

Schnieke, A.E. and others. 1997. Human Factor IX transgenic sheep produced by transfer of nuclei from transfected fetal fibroblasts. *Science* 278:2130–2133.

Wall, R.J. 1999. Biotechnology for the production of modified and innovative animal products: transgenic livestock bioreactors. *Livestock Prod. Sci.* 59:243–255.

Wilmut, I., K. Campbell, and C. Tudge. 1997. *The Second Creation*. Farrar, Straus, and Giroux, New York.

Vaccinating for Animal Health

Kwang, J. 2000. Fishing for vaccines. *Nature Biotech.* 18:1145–1146.

Lorenzen, N. and others. 2000. Immunoprophylaxis in fish by injection of mouse antibody genes. *Nature Biotech.* 18:1177–1180.

Wolff, J.A. and others. 1990. Direct gene transfer into mouse muscle in vivo. *Science* 237:1465–1468.

Engineering Foods for Animals

Mellon, M., C. Benbrook, and K.L. Benbrook. 2001. *Hogging It: Estimates of Antimicrobial Abuse in Livestock*. Union of Concerned Scientists Publication, Cambridge, MA.

Palumbi, S.R. 2001. *The Evolution Explosion: How Humans Cause Rapid Evolutionary Change*. W.W. Norton, New York.

von Wettstein, D., G. Mikhaylenko, J.A. Froseth, and C.G. Kannangara. 2000. Improved barley broiler feed with transgenic malt containing heat-stable (1,3-1,4)-β-glucanase. *Proc. Natl. Acad. Sci. USA* 97:13512–13517.

Cloned Organ-donor Pigs

Betthauser, J. and others. 2000. Production of cloned pigs from *in vitro* systems. *Nature Biotech.* 18:1055–1059.

Dai, Y. and others. 2002. Targeted disruption of the α1,3-galactosyltransferase gene in cloned pigs. *Nature Biotech.* 20:251–255.

Hooper, E. 1999. *The River: A Journey to the Source of HIV and AIDS*. Little Brown, New York.

Li, G.-P. and others. 2000. Cloned piglets born after nuclear transplantation of embryonic blastomeres into porcine oocytes matured *in vivo*. *Cloning* 2:45–52.

Onishi, A. and others. 2000. Pig cloning by microinjection of fetal fibroblast nuclei. *Science* 289:1188–1190.

Paradis, K. and others. 1999. Search for cross-species transmission of porcine endogenous retrovirus in patients treated with living pig tissue. *Science* 285:1236–1241.

Platt, J.L. 2000. Xenotransplantation: new risks, new gains. *Nature* 407:27–30.
Polejaeva, I.A. and others. 2000. Cloned pigs produced by nuclear transfer from adult somatic cells. *Nature* 407:86–90.
van der Laan, L.J.W. and others. 2000. Infection by orcine endogenous retrovirus after islet xenotransplantation in SCID mice. *Nature* 407:90–94.

Possibilities with Poultry

Bosselman, R. and others. 1989. Germline transmission of exogenous genes in the chicken. *Science* 243:533–535.
Graves, A. 2001. Clone farm. *New Scientist* 171:4–5.
Harvey, A.J., G. Speksnijder, L.R. Baugh, J.A. Morris, and R. Ivarie. 2002. Consistent production of transgenic chickens using replication-deficient retroviral vectors and high-throughput screening procedures. *Poultry Sci.* 81:202–212.
Harvey, A.J., G. Speksnijder, L.R. Baugh, J.A. Morris, and R. Ivarie. 2002. Expression of exogenous protein in the egg white of transgenic chickens. *Nature Biotech.* 19:396–399.
Love, J., C. Gribbin, C. Mather, and H. Sang. 1994. Transgenic birds by DNA microinjection. *Biotechnology* 12:60–63.
Salter, D.W. and others. 1986. Gene insertion into the chicken germ line by retroviruses. *Poultry Sci.* 65:1455–1458.
Sang, H. 1994. Transgenic chickens—methods and potential applications. *Trends Biotech.* 12:415–419.
Sherman, A. and others. 1998. Transposition of the *Drosophila* element *mariner* into the chicken germ line. *Nature Biotech.* 16:1050–1053.

Copy Cats

Chesné, P., P.G. Adenot, C. Viglietta, M. Baratte, L. Boulanger, and J.-P. Renard. 2002. Cloned rabbits produced by nuclear transfer from adult somatic cells. *Nature Biotech.* 20:366–369.
Holden, C. 2002. Carbon-copy clone is the real thing. *Science* 295:1443–1444.
Shin, T. and others. 2002. A cat cloned by nuclear transplantation. *Nature* 415:859.

Good-bye Dolly

Brem, G., and B. Kühholzer. 2002. The recent history of somatic cell cloning in mammals. *Cloning and Stem Cells* 4:57–63.
Cibelli, J.B., K.H. Campbell, G.E. Seidel, M.D. West, and R.P. Lanza. 2002. The health profile of cloned animals. *Nature Biotech.* 20:13–14.
Hill, J.R. and others. 1999. Clinical and pathological features of cloned transgenic calves and fetuses (13 case studies). *Theriogenology* 51:1451–1465.
Humphreys, D. and others. 2001. Epigenetic instability in ES cells and cloned mice. *Science* 293:95–97.
Ogonuki, N. and others. 2002. Early death of mice cloned from somatic cells. *Nature Genet.* 30:253–254.
Shields, P.G. and others. 1999. Analysis of telomere lengths in cloned sheep. *Nature* 399:316–317.

Xeu, F. and others. 2002. Aberrant patterns of X chromosomal inactivation in bovine clones. *Nature Genet.* 31:216–220.

6 Fields, Forests, and Streams

Pulp Nonfiction

Dinus, R.J., P. Payne, M.M. Sewell, V.L. Chiang, and G.A. Tuskan. 2001. Genetic modification of short rotation popular wood: properties for ethanol fuel and fiber productions. *Crit. Rev. Plant Sci.* 20:51–69.

Hu, W.-J. and others. 1999. Repression of lignin biosynthesis promotes cellulose accumulation and growth in transgenic trees. *Nature Biotech.* 17:808–812.

Lapierre, C. and others. 1999. Structural alterations of lignins in transgenic poplars with depressed cinnamyl alcohol dehydrogenase or caffeic acid *O*-methyltransferase activity have an opposite impact on the efficiency of industrial Kraft pulping. *Plant Physiol.* 119:153–163.

Mann, C.C., and M.L. Plummer. 2002. Forest biotech edges out of the lab. *Science* 295:1626–1629.

Merkle, S.A., and J.F.D. Dean. 2000. Forest tree biotechnology. *Curr. Opin. Biotechnol.* 11:298–302.

National Research Council. 2002. *Environmental Effects of Transgenic Plants.* National Academy Press, Washington, DC.

Pilate, G. and others. 2002. Field and pulping performances of transgenic trees with altered lignification. *Nature Biotech.* 20:607–612.

Antimalarial Mosquitoes

Capurro, M, J. and others. 2000. Virus-expressed, recombinant single-chain antibody blocks sporozoite infection of salivary glands in *Plasmodium gallinaceum*-infected *Aedes aegypti. Am. J. Tropical Med. Hyg.* 62:427–433.

Catteruccla, F. and others. 2000. Stable germline transformation of the malaria mosquito *Anopheles stephensi. Nature* 405:959–962.

Clarke, T. 2002. Mosquitoes minus malaria. *Nature* 419:429–430.

Coates, C.J. 2000. A mosquito transformed. *Nature* 405:900–901.

D'Antonio, M. 2001. Making a new mosquito. *Discover* 22(5):64–69.

Enserink, M. 2002. Ecologists see flaws in transgenic mosquito. *Science* 297:30–31.

Ghosh, A., M.J. Edwards, and M. Jacobs-Lorena. 2000. The journey of the malaria parasite in the mosquito: hopes for a new century. *Parasitol. Today* 16:196–201.

Ito, J., A. Ghosh, L.A. Moreira, E.A. Wimmer, and M. Jacobs-Lorena. 2002. Transgenic anopheline mosquitoes impaired in transmission of a malaria parasite. *Nature* 417:452–455.

James, A.A. and others. 1999. Controlling malaria transmission with genetically-engineered, *Plasmodium*-resistant mosquitoes: milestones in a model system. *Parassitologia* 41:461–471.

Jasinskiene, N. and others. 1998. Stable transformation of the yellow fever mosquito, *Aedes aegypti*, with the *Hermes* element from the housefly. *Proc. Natl. Acad. Sci. USA* 95:3743–3747.

Kidwell, M.G., and A.R. Wattam. 1998. An important step forward in the genetic manipulation of mosquito vectors in human disease. *Proc. Natl. Acad. Sci. USA* 95:3349–3350.

O'Brochta, D.A., and P.W. Atkinson. 1998. Building the better bug. *Sci. Am.* 279(6):90–95.

Fat, Sexy Salmon

Allendorf, F.W., and R.S. Waples. 1996. Conservation and genetics of salmonid fishes. Pp. 238–280 in: *Conservation Genetics: Case Histories from Nature*, J.C. Avise and J.L. Hamrick, editors. Chapman & Hall, New York.

Devlin, R.H. 1997. Transgenic salmonids. Pp. 105–117 in: *Transgenic Animals: Generation and Use*, L.M. Houdebine, editor. Harwood Academic, Amsterdam.

Devlin, R.H., T.Y. Yesaki, C.A. Blagl, E.M. Donaldson, P. Swanson, and W.-K. Chan. 1994. Extraordinary salmon growth. *Nature* 371:209–210.

Food and Agricultural Organization. 2000. *The State of World Fisheries and Aquaculture 2000*. FAO, Rome.

Hedrick, P.W. 2001. Invasion of transgenes from salmon or other genetically modified organisms into natural populations. *Can. J. Fish. Aquat. Sci.* 58:841–844.

Kapuchinski, A.R., and E.M. Hallerman. 1991. Implications of introduction of transgenic fish into natural ecosystems. *Can. J. Fish. Aquat. Sci.* 48:99–107.

Knibb, W. 1997. Risk from genetically engineered and modified marine fish. *Transgenic Res.* 6:59–67.

McDowell, N. 2002. Stream of escaped farm fish raises fears for wild salmon. *Nature* 416:571.

Muir, W.M., and R.D. Howard. 1999. Possible ecological risks of transgenic organism release when transgenes affect mating success: sexual selection and the Trojan horse hypothesis. *Proc. Natl. Acad. Sci. USA* 13853–13856.

National Research Council. 1996. *Upstream: Salmon and Society in the Pacific Northwest.* National Academy Press, Washington, DC.

Pew Foundation. 2003. *Future Fish: Issues in Science and Regulation of Transgenic Fish.* Pew Initiative on Food and Biotechnology, Washington, DC.

Reichhardt, T. 2000. Will souped up salmon sink or swim? *Nature* 406:10–12.

Antifreeze Proteins

Chen, L., A.L. DeVries, and C.-H.C. Cheng. 1997. Evolution of antifreeze glycoprotein gene from a trypsinogen gene in Antarctic notothenioid fish. *Proc. Natl. Acad. Sci. USA* 94:3811–3816.

di Prisco, G., and E. Pisano, editors. 1998. *Evolution of the Antarctic Ichthyofauna.* Springer-Verlag, New York.

Fletcher, G.L., S.V. Goddard, and Y. Wu. 1999. Antifreeze proteins and their genes: from basic research to business opportunity. *Chemtech* 30(6):17–28.

Graham, L.A., Y.-C. Liou, V.K. Walker, and P.L. Davies. 1997. Hyperactive antifreeze protein from beetles. *Nature* 388:727–728.

Hightower, R., C. Baden, E. Penzes, P. Lund, and P. Dunsmuir. 1991. Expression of antifreeze proteins in transgenic plants. *Plant Mol. Biol.* 17:1013–1021.

Logsdon, J.M. Jr., and W.F. Doolittle. 1997. Origin of antifreeze protein genes: a cool tale in molecular evolution. *Proc. Natl. Acad. Sci. USA* 94:3485–3487.

Pandian, T.J. 2001. Guidelines for research and utilization of genetically modified fish. *Current Sci.* 81:1172–1178.

Mutation-Detecting Fish

Amanuma, K., H. Takeda, H. Amanuma, and Y. Aoki. 2000. Transgenic zebrafish for detecting mutations caused by compounds in aquatic environments. *Nature Biotech.* 18:62–65.

Battalora, M., and R. Tennant. 1999. The use of transgenic mice in mutagenesis and carcinogenesis research and in chemical safety assessment. Pp. 111–126 in: *Molecular Biology of the Toxic Response*, A. Puga and K. Wallace, editors. Taylor & Francis, Philadelphia.

Winn, R.N. 2001. Transgenic fish as models in environmental toxicology. *ILAR J.* 42:322–329.

Winn, R.N., M. Norris, S. Muller, C. Torres, and K. Brayer. 2001. Bacteriophage λ and plasmid pUR288 transgenic fish models for detecting *in vivo* mutations. 2001. *Marine Biotech.* 3:S185–S195.

Sentinels of Aquatic Pollution

Carvan M.J. III, T.P. Dalton, G.W. Stuart, and D.W. Nebert. 2000. Transgenic zebrafish as sentinels for aquatic pollution. *Ann. N.Y. Acad. Sci.* 919:133–147.

Gibbs, P.D.L., A. Gray, and G. Thorgaard. 1994. Inheritance of P element and reporter gene sequences in zebrafish. *Mol. Marine Biol. Biotech.* 3:317–326.

Maclean, N. 1998. Regulation and exploitation of transgenes in fish. *Mutation Res.* 399:255–266.

Powers, D.A. 1989. Fish as model systems. *Science* 246:352–358.

Transgenic Environmental Biosensors

Alloway, B.J., and D.C. Ayres. 1993. *Chemical Principles of Environmental Pollution.* Chapman & Hall, New York.

Burlage, R.S. 1999. Green fluorescent bacteria for the detection of landmines in a minefield. *Second Int. Symp. on GFPs,* San Diego, CA.

Coghlan, A. 1999. Bugs give clean sites a glowing recommendation. *New Scientist* 164(2215):20.

Knasmuller, S. and others. 1998. Detection of genotoxic effects of heavy metal contaminated soils with plant bioassays. *Mutat. Res.* 420:37–48.

Kovalchuk, I., O. Kovalchuk, A. Arkhipov, and B. Hohn. 1998. Transgenic plants are sensitive bioindicators of nuclear pollution caused by the Chernobyl accident. *Nature Biotech.* 16:1054–1059.

Kovalchuk, O., V. Titov, B. Hohn, and I. Kovalchuk. 2001. A sensitive transgenic plant system to detect toxic inorganic compounds in the environment. *Nature Biotech.* 19:568–572.

Phytoremediation of Mercury Poisons

Cai, X.-H., J. Adhiya, S. Traina, and R. Sayre. 1998. Heavy metal binding properties of wild type and transgenic algae (*Chlamydomonas* sp.). Pp. 189–192 in: *New Developments in Marine Biotechnology*, L. Le Gal and H.O. Halvorson, editors. Plenum Press, New York.

Doucleff M., and N. Terry. 2002. Pumping out the arsenic. *Nature Biotech.* 20:1094–1095.

Ensley, B., and I. Raskin, editors. 1999. *Phytoremediation of Toxic Metals: Using Plants to Clean-up the Environment.* John Wiley & Sons, New York.

Harada, M. 1995. Minamata disease: methylmercury poisoning in Japan caused by environmental pollution. *Crit. Rev. Toxicol.* 25:1–24.

Heaton, A.C.P., C.L. Rugh, N.J. Wang, and R.B. Meagher. 1998. Phytoremediation of mercury and methylmercury polluted soils using genetically engineered plants. *J. Soil Contamination* 7:497–509.

Salt, D.E., R.D. Smith, and I. Raskin. 1998. Phytoremediation. *Annu. Rev. Plant Physiol. Plant Mol. Biol.* 49:643–668.

Terry, N., and G. Bañuelos, editors. 2000. *Phytoremediation of Contaminated Soil and Water.* Lewis Publishers, Boca Raton, FL.

Phytoremediation of Organic Pollutants

Anderson, T.A., E.A. Guthrie, and B.T. Walton. 1993. Bioremediation in the rhizosphere: plant roots and associated microbes clean contaminated soil. *Environ. Sci. Tech.* 27:2630–2636.

Best, E.P., M.E. Zappi, H.L. Fredrickson, S.L. Sprecher, S.L. Larson, and M. Ochman. 1997. Screening of aquatic and wetland plant species for phytoremediation of explosives-contaminated groundwater from the Iowa Army Ammunition Plant. *Ann. N.Y. Acad. Sci.* 829:179–194.

French, C.E., S.J. Rosser, G.J. Davies, S. Nicklin, and N.C. Bruce. 1999. Biodegradation of explosives by transgenic plants expressing pentaerythritol tetranitrate reductase. *Nature Biotech.* 17:491–494.

Meagher, R.B. 2000. Phytoremediation of toxic elemental and organic pollutants. *Curr. Opin. Plant Biol.* 3:153–162.

Salt and Drought Plants

Kasuga, M., Q. Liu, S. Miura, K. Yamaguchi-Shinozaki, and K. Shinozaki. 1999. Improving plant drought, salt, and freezing tolerance by gene transfer of a single stress-inducible transcription factor. *Nature Biotech.* 17:287–291.

Mitra, J. 2001. Genetics and genetic improvement of drought resistance in crop plants. *Current Sci.* 80:758–763.

Moffat, A.S. 2002. Finding new ways to protect drought-stricken plants. *Science* 296:1226–1229.

Somerville, C., and J. Briscoe. 2001. Genetic engineering and water. *Science* 292:2217.

Zhang, H.-X., and E. Blumwald. 2001. Transgenic salt-tolerant tomato plants accumulate salt in foliage but not in fruit. *Nature Biotech.* 19:765–768.

Zhu, J.-K. 2001. Plant salt tolerance. *Trends Plant Sci.* 6:66–71.

Bioremediating Bacteria

Brim, H. and others. 2000. Engineering *Deinococcus radiodurans* for metal remediation in radioactive mixed waste environments. *Nature Biotech.* 18:85–90.

Daly, M.J. 2000. Engineering radiation-resistant bacteria for environmental biotechnology. *Curr. Opin. Biotech.* 11:280–285.

Guengerich, F.P. 1995. Cytochrome P450 proteins and potential utilization in biodegradation. *Environ. Health Perspect.* 103:25–28.

Iranzo, M., I. Sainz-Pardo, R. Boluda, J. Sánchez, and S. Mormeneo. 2001. The use of microorganisms in environmental remediation. *Ann. Microbiol.* 51:135–143.

Kellner, D.G., S.A. Maves, and S.G. Sligar. 1997. Engineering cytochrome P450s for bioremediation. *Curr. Opin. Biotech.* 8:274–278.

Pieper, D.H., and W. Reineke. 2000. Engineering bacteria for bioremediation. *Curr. Opin. Biotech.* 11:262–270.

Cries over Spilled Oil

Atlas, R.M. 1981. Microbial degradation of petroleum hydrocarbons: an environmental perspective. *Microbiol. Rev.* 45:180–209.

Kapley, A., H.J. Purohit, S. Chhatre, R. Shanker, T. Chakrabarti, and P. Khanna. 1999. Osmotolerance and hydrocarbon degradation by a genetically engineered microbial consortium. *Bioresource Technol.* 67:241–245.

Prince, R.C. 1993. Petroleum spill bioremediation in marine environments. *Crit. Rev. Microbiol.* 19:211–242.

Sikdar, S.K., R.L. Irvine, and P.N. Lancaster, editors. 1998. *Bioremediation: Principles and Practice.* Technomic, Lancaster, PA.

Rabbit Contraception

Angulo, E., and B. Cooke. 2002. First synthesize new viruses then regulate their release? The case of the wild rabbit. *Mol. Ecol.* 11:2691–2702.

Cowan, P.E., and C.H. Tyndale-Biscoe. 1997. Australian and New Zealand mammal species considered to be pests or problems. *Reprod. Fertil. Devel.* 9:27–36.

Fenner, F., and B. Fantini. 1999. *Biological Control of Vertebrate Pests: The History of Myxomatosis, an Experiment in Evolution.* CABI, New York.

Hood, G.M., P. Chesson, and R.P. Pech. 2000. Biological control using sterilizing viruses: host suppression and competition between viruses in non-spatial models. *J. Appl. Ecol.* 37:914–925.

Kerr, P.J. and others. 1999. Infertility in female rabbits (*Oryctolagus ciniculus*) alloimmunized with the rabbit zona pellucida protein ZPB either as a purified recombinant protein or expressed by recombinant myxoma virus. *Biol. Reprod.* 61:606–613.

McCallum, H. 1996. Immunocontraception for wildlife population control. *Trends Ecol. Evol.* 11:491–493.

Morell, V. 1993. Australian pest control by virus causes concern. *Science* 261:683–684.

Williams, K., I. Parer, B. Coman, J. Burley, and M. Braysher. 1995. *Managing Vertebrate Pests: Rabbits.* Australian Govt. Public Service, Canberra.

Daughterless Carp

National Research Council. 1995. *Understanding Marine Biodiversity: A Research Agenda for the Nation*. National Academy Press, Washington, DC.

National Research Council. 1996. *Stemming the Tide: Controlling Introductions of Non-indigenous Species by Ships' Ballast Water*. National Academy Press, Washington, DC.

Naylor, R.L., S.L. Williams, and D.R. Strong. 2001. Aquaculture—a gateway for exotic species. *Science* 294:1655–1656.

Nowak, R. 2002. Gene warfare. *New Scientist* 174(2342):6.

Utter, F.M., O.W. Johnson, G.H. Thorgaard, and P.S. Rabinovitch. 1983. Measurement and potential applications of induced triploidy in Pacific salmon. *Aquaculture* 35:125–135.

Pesticide Detoxification

Calver, M.C., and D.R. King. 1986. Controlling vertebrate pests with fluoracetate: lessons in wildlife management, bioethics, and coevolution. *J. Biol. Edu.* 20:257–262.

Gregg, K., B. Hamdorf, K. Henderson, J. Kopency, and C. Wong. 1998. Genetically modified ruminal bacteria protect sheep from fluoroacetate poisoning. *Applied Environ. Microbiol.* 64:3496–3498.

Mead, R.J., A.J. Oliver, D.R. King, and P.H. Hubach. 1985. The co-evolutionary role of fluoroacetate in plant animal interaction in Australia. *Oikos* 44:55–60.

Twigg, L.E., and D.R. King. 2000. Artificially enhanced tolerance to fluoroacetate and its implications for wildlife conservation. *Pacific Conserv. Biol.* 6:9–13.

Blue Rose Petals and a Mauve Carnation

Aida, R., S. Kishimoto, Y. Tanaka, and M. Shibata. 2000. Modification of flower color in torenia (*Torenia fournieri* Lind.) by genetic transformation. *Plant Sci.* 153:33–42.

Brown, K. 2002. Something to sniff at: unbottling floral scent. *Science* 296:2327–2329.

deVetten, N., J. terHorst, H.P. vanSchaik, A. deBoer, J. Mol, and R.A. Koes. 1999. A cytochrome b(5) is required for full activity of flavonoid 3',5'-hydroxylase, a cytochrome P450 involved in the formation of blue flower colors. *Proc. Natl. Acad. Sci. USA* 96:778–783.

Mercuri, A., A. Saccheti, L. De Benedetti, T. Schiva, and S. Alberti. 2001. Green fluorescent flowers. *Plant Sci.* 161:961–968.

Meyer, P., I. Heidmann, G. Forkmann, and H. Saedler. 1987. A new petunia flower colour generated by transformation of a mutant with a maize gene. *Nature* 330:677–678.

Mol, J., E. Cornish, J. Mason, and R. Koes. 1999. Novel coloured flowers. *Curr. Opin. Biotech.* 10:198–201.

Shimada, Y., M. Ohbayashi, R. Nakano-Shimada, Y. Okinaka, S. Kiyokawa, and Y. Kikuchi. 2001. Genetic engineering of the anthocyanin biosynthetic

pathway with flavenoid-3',5'-hydroxylase: specific switching of the pathway in petunia. *Plant Cell Rep.* 20:456–462.
Wright, K. 2001. Species on ice. *Discover* 22(9):28–29.

No-mow Lawns

Bacon, C.W., M.D. Richardson, and J.F. White, Jr. 1997. Modification and uses of endophyte-enhanced turfgrasses: a role for molecular technology. *Crop Sci.* 37:1415–1425.
Duble, R.L. 1996. *Turfgrasses: Their Management and Use in the Southern Zone*, 2nd ed. Texas A&M University Press, College Station.
Neff, M.M. and others. 1999. *BAS1*: A gene regulating brassinosteroid levels and light responsiveness in *Arabidopsis*. *Proc. Natl. Acad. Sci. USA* 96:15316–15323.

Sperm Whale Oils and Jojoba Waxes

Alen, K.R. 1980. *Conservation and Management of Whales*. University of Washington Press, Seattle.
Brooks, W.H. 1978. Jojoba—a North American desert shrub; its ecology, possible commercialization, and potential as an introduction into other arid regions. *J. Arid Environ.* 1:227–236.
Brouin, P., S. Gettner, and C. Somerville. 1999. Genetic engineering of plant lipids. *Annu. Rev. Nutr.* 19:197–216.
Lardizabal, K.D., J.G. Metz, T. Sakamoto, W.C. Hutton, M.R. Pollard, and M.W. Lassner. 2000. Purification of a jojoba embryo wax synthase, cloning of its cDNA, and production of high levels of wax in seeds of transgenic *Arabidopsis*. *Plant Physiol.* 122:645–655.
Tonnessen, J.N., and A.O. Johnsen. 1982. *The History of Modern Whaling*. University of California Press, Berkeley.

Rescuing Endangered Species

Corley-Smith, G.E., and B.P. Brandhorst. 1999. Preservation of endangered species and populations: a role for genome banking, somatic cloning, and androgenesis. *Mol. Reprod. Dev.* 53:363–367.
Frankham, R., J.D. Ballou, and D.A. Briscoe. 2002. *Introduction to Conservation Genetics*. Cambridge University Press, Cambridge.
Holden, C., 2002. Reviving the Tasmanian tiger. *Science* 296:1797.
Lanza, R.P. and others. 2000. Cloning of an endangered species (*Bos gaurus*) using interspecies nuclear transfer. *Cloning* 2:79–90.
Lanza, R.P., B.L. Dresser, and P. Damiani. 2000. Cloning Noah's ark. *Sci. Am.* 283(5):84–89.
Lee, K.-Y., H. Huang, B. Ju, Z. Yang, and S. Lin. 2002. Cloned zebrafish by nuclear transfer from long-term-cultured cells. *Nature Biotech.* 20:795–799.
Loi, P., G. Ptak, B. Barboni, J. Fulka, Jr., P. Cappai, and M. Clinton. 2001.

Genetic rescue of an endangered mammal by cross-species nuclear transfer using post-mortem somatic cells. *Nature Biotech.* 19:962–964.
Ryder, O.A., A. McLaren, S. Brenner, Y.P. Zhang, and K. Benirschke. 2000. DNA banks for endangered species. *Science* 288:275–277.

Concluding Thoughts

Hardin, G. 1968. The tragedy of the commons. *Science* 162:1243–1248.
National Research Council. 2002. *Animal Biotechnology: Science-Based Concerns.* National Academy Press, Washington, DC.
Sayler, G.S., and S. Ripp. 2000. Field applications of genetically engineered organisms for bioremediation processes. *Curr. Opin. Biotech.* 11:286–289.

7 Genetic Tinkering with Humans

Fukuyama, F. 2002. *Our Posthuman Future: Consequences of the Biotechnology Revolution.* Farrar, Straus and Giroux, New York.
San, L.P., and E.P.H. Yap, editors. 2001. *Frontiers in Human Genetics: Diseases and Technologies.* World Scientific, Singapore.
Stock, G. 2002. *Redesigning Humans: Our Inevitable Genetic Future.* Houghton Mifflin, Boston.

Gene Therapies on SCIDs

Aiuti, A. and others. 2002 Correction of ADA-SCID by stem cell gene therapy combined with nonmyeloablative conditioning. *Science* 296:2410–2413.
Anderson, W.F. 1995. Gene therapy. *Sci. Am.* 273(3):124–128.
Anderson, W.F. 2000. The best of times, the worst of times. *Science* 288:627–630.
Cavazzana-Calvo, M. and others. 2000. Gene therapy of human combined immunodeficiency (SCID)-XI disease. *Science* 288:669–672.
Check, E. 2002. A tragic setback. *Nature* 420:116–118.
Crystal, R.G. 1995. Transfer of genes to humans: early lessons and obstacles to success. *Science* 270:404–410.
Lyon, J., and P. Gorner. 1995. *Altered Fates: Gene Therapy and the Retooling of Human Life.* W.W. Norton, New York.
Mulligan, R.C. 1993. The basic science of gene therapy. *Science* 260:926–932.

Gene Therapies in the Works

Boyce, N. 2000. Engineering a cure. *New Scientist* 164(2218):6.
Friedmann, T., editor. 1999. *The Development of Human Gene Therapy.* Cold Spring Harbor Laboratory Press, Cold Spring Harbor, NY.
Kang, R., S.C. Ghivizzani, T.S. Muzzonigro, J.H. Herndon, P.D. Robbins, and C.H. Evans. 2000. Orthopaedic applications of gene therapy—from concept to clinic. *Clin. Orthopaed. Related Res.* 375:324–337.
Lattime, E.C., and S.L. Gerson, editors. 2002. *Gene Therapy of Cancer.* Academic Press, San Diego, CA.
Olefsky, J.F. 2000. Gene therapy for rats and mice. *Nature* 408:420–421.

Simon, E.J. 2002. Human gene therapy: genes without frontiers? *Am. Biol. Teacher* 64(4):264–270.

Zanjani, E.D., and W.F. Anderson. 1999. Prospects for *in utero* human gene therapy. *Science* 285:2084–2088.

New Angles on Gene Therapy Vectors

Ferber, D. 2001. Gene therapy: safer and virus-free? *Science* 294:1638–1642.

Kagawa, Y., Y. Inoki, and H. Endo. 2001. Gene therapy by mitochondrial transfer. *Adv. Drug Delivery Rev.* 49:107–119.

Marshall, E. 2000. Gene therapy on trial. *Science* 288:951–957.

Pfeifer, A., and I.M. Verma. 2001. Gene therapy: promises and problems. *Annu. Rev. Genomics Hum. Genet.* 2:177–211.

Tissue Therapy via Gene Therapy: The Angiogenesis Story

Abbott, A. 2001. Genetic medicine gets real. *Nature* 411:410–412.

Henry, T.D. and others. 1999. Final results of the VIVA trial of rhVEGF for human therapeutic angiogenesis. *Circulation* 100:I-476 (suppl. 1).

Isner, J.M. and others. 1995. Arterial gene therapy for therapeutic angiogenesis in patients with peripheral artery disease. *Circulation* 91:2687–2692.

Losordo, D.W., P.R. Vale, and J.M. Isner. 1999. Gene therapy for myocardial angiogenesis. *Am. Heart J.* 138:5132–5141.

Moseley, J.B. and others. 2002. A controlled trial of arthroscopic surgery for osteoarthritis of the knee. *N. Engl. J. Med.* 347:81–88.

Rosengart, T.K. and others. 1999. Six-month assessment of phase I trial of angiogenic gene therapy for the treatment of coronary artery disease using direct intramyocardial administration of an adenovirus vector expressing the VEGF121 cDNA. *Ann. Surg.* 100:466–470.

Vale, P.R. and others. 2000. Left ventricular electromechanical mapping to assess efficacy of phVEGF165 gene transfer for therapeutic angiogenesis in chronic myocardial ischemia. *Circulation* 102:965–974.

Wolf, P. and others. 2000. Topical treatment with liposomes containing T4 endonuclease V protects human skin *in vivo* from ultraviolet-induced upregulation of interleukin-10 and tumor necrosis factor-α. *J. Invest. Dermatol.* 114:149–156.

Tissue Therapy via Therapeutic Cloning

Holden, C., and G. Vogel. 2002. Plasticity: time for a reappraisal? *Science* 296:2126–2129.

Lanza, R.P., J.B. Cibelli, and M.D. West. 1999. Prospects for the use of nuclear transfer in human transplantation. *Nature Biotech.* 17:1171–1174.

National Academy of Sciences. 2001. *Stem Cells and the Future of Regenerative Medicine*. National Academy Press, Washington, DC.

Vogelstein, B., B. Alberts, and K. Shine. 2002. Please don't call it cloning! *Science* 295:1237.

Embryonic Stem Cells

Bonetta, L. 2001. Storm in a culture dish. *Nature* 413:345–346.

Green, R.M. 2001. *The Human Embryo Research Debates: Bioethics in the Vortex of Controversy*. Oxford University Press, New York.

Holland, S., K. Lebacqz, and L. Zoloth, editors. 2001. *The Human Embryonic Stem Cell Debate: Science, Ethics, and Public Policy*. The MIT Press, Cambridge, MA.

Kass, L.R., and J.Q. Wilson. 1998. *The Ethics of Human Cloning*. AEI Press, Washington, DC.

Kaufman, D.S., E.T. Hanson, R.L. Lewis, R. Auerbach, and J.A. Thompson. 2001. Hematopoietic colony-forming cells derived from human embryonic stem cells. *Proc. Natl. Acad. Sci. USA* 98:10716–10721.

Lanza, R.P., J.B. Cibelli, and M.D. West. 1999. Human therapeutic cloning. *Nature Med.* 5:975–977.

Marshak, D.R., R.L. Gardner, and D. Gottlieb, editors. 2001. *Stem Cell Biology*. Cold Spring Harbor Laboratory Press, Cold Spring Harbor, NY.

McLaren, A. 2001. Ethical and social considerations of stem cell research. *Nature* 414:129–131.

More on Stem Cells

Abbott, A., and D. Cyranoski. 2001. China plans "hybrid" embryonic stem cells. *Nature* 413:339.

Ainsworth, C. 2001. Warning light. *New Scientist* 170(2297):13.

Cibelli, J.B. and others. 2002. Parthenogenetic stem cells in nonhuman primates. *Science* 295:819.

DeWitt, N., editor. 2001. Stem cells. *Nature* 414:87–131.

Edwards, R.G. 2001. IVF and the history of stem cells. *Nature* 413:349–351.

Klug, M.G., M.H. Soonpaa, G.Y. Kon, and L.J. Field. 1996. Genetically selected cardiomyocytes from differentiating embryonic stem cells form stable intracardiac grafts. *J. Clin. Invest.* 98:216–224.

McKay, R. 2000. Stem cells—hype and hope. *Nature* 406:361–364.

Solter, D., and J. Gearhart. 1999. Putting stem cells to work. *Science* 283:1468–1470.

Thomson, J.A. and others. 1998. Embryonic stem cell lines derived from human blastocysts. *Science* 282:1145–1147.

Trounson, A. 2002. The genesis of embryonic stem cells. *Nature Biotech.* 20:237–238.

Vogel, G. 2002. Studies cast doubt on plasticity of adult cells. *Science* 295:1989–1920.

Whole-Human Clones

Brock, D.W. 2002. Human cloning and our sense of self. *Science* 296:314–316.

Cibelli, J.B., A.A. Kiessling, K. Cunniff, C. Richards, R.P. Lanza, and M.D. West. 2001. Somatic cell nuclear transfer in humans: pronuclear and early embryonic development. *J. Regenerative Med.* 2:25–31.

Cibelli, J.B., R.P. Lanza, and M.D. West. 2002. The first human cloned. *Sci. Am.* 286(1)44–51.

National Academy of Sciences. 2002. *Human Reproductive Cloning*. National Academy Press, Washington, DC.

Schatten, G., R. Prather, and I. Wilmut. 2003. Cloning claim is science fiction, not science. *Science* 299:344.

Engineering the Germline

Brinster, R.L. 2002. Germline stem cell transplantation and transgenesis. *Science* 296:2174–2176.

Chan, A.W.S., K.Y. Chong, C. Martinovich, C. Simerly, and G. Schatten. 2001. Transgenic monkeys produced by retroviral gene transfer into mature oocytes. *Science* 291:309–312.

Kevles, D.J. 1995. *In the Name of Eugenics*. Harvard University Press, Cambridge, MA.

Knight, J. 2001. Biology's last taboo. *Nature* 413:12–15.

Nagano, M., T. Shinohara, M.R. Avarbock, and R.L. Brinster. 2000. Retrovirus-mediated gene delivery into male germ line cells. *FEBS Lett.* 475:7–10.

Stock, G., and J. Campbell, editors. 2000. *Engineering the Human Germline*. Oxford University Press, New York.

More New-Age Eugenics

Wade, N. 2001. *Life Script: How the Human Genome Discoveries Will Transform Medicine and Enhance Your Health*. Simon & Schuster, New York.

Appendix: Tools and Workshops of Genetic Engineering

Restricted Activities

Boyer, H.W. 1971. DNA restriction and modification mechanisms in bacteria. *Annu. Rev. Microbiol.* 25:153–176.

Hedgpeth, J., M.H. Goodman, and H.M. Boyer. 1972. DNA nucleotide sequence restricted by the RI endonuclease. *Proc. Natl. Acad. Sci. USA* 69:3448–3452.

Linn, S., and W. Arber. 1968. Host specificity of DNA produced by *Escherichia coli*, X. *In vitro* restriction of phage fd replicative form. *Proc. Natl. Acad. Sci. USA* 59:1300–1306.

Meselson, M., and R. Yuan. 1968. DNA restriction enzyme from *E. coli*. *Nature* 217:1110–1114.

Cut and Paste

Kessler, C. 1987. Class II restriction endonucleases. Pp. 225–279 in: *Cytogenetics*, G. Obe and A. Basler, editors. Springer-Verlag, Berlin.

Roberts, J.R. 1984. Restriction and modification enzymes and their recognition sequences. *Nucleic Acids Res.* 12:167–204.

Copy and Duplicate

Cohen, S.N., A.C.Y. Chang, H.W. Boyer, and R.B. Helling. 1973. Construction of biologically functional bacterial plasmids *in vitro*. *Proc. Natl. Acad. Sci. USA* 70:3240–3244.

Maniatis, T.E., E.F. Fritsch, and J. Sambrook. 1982. *Molecular Cloning: A Laboratory Manual.* Cold Spring Harbor Laboratory Press, Cold Spring Harbor, NY.

Viral Vectors

Bushman, F. 2002. *Lateral DNA Transfer: Mechanisms and Consequences.* Cold Spring Harbor Laboratory Press, Cold Spring Harbor, NY.

Jackson, D., R. Symons, and P. Berg. 1972. Biochemical method for inserting new genetic information into DNA of simian virus 40: circular SV40 DNA molecules containing lambda phage genes and the galactose operon of *Escherichia coli. Proc. Natl. Acad. Sci. USA* 69:2904–2909.

Marchant, J. 2000. Could a gene cocktail halt Parkinson's? *New Scientist* 168(2261):12.

Phillips, M.I., editor. 2002. Gene therapy methods. *Meth. Enzymol.* 346.

Wang, B., J. Li, and X. Xiao. 2000. Adeno-associated virus vector carrying human minidystrophin genes effectively ameliorates muscular dystrophy in *mdx* mouse model. *Proc. Natl. Acad. Sci. USA* 97:13714–13719.

Jumping Genes

Izsvak, Z., S. Ivics, and R.H. Plasterk. 2000. Sleeping Beauty, a wide host-range transposon vector for genetic transformation in vertebrates. *J. Mol. Biol.* 302:93–102.

Kidwell, M.G. 1993. Lateral transfer in natural populations of eukaryotes. *Annu. Rev. Genet.* 27:235–256.

Linden, R.M. 2002. Gene therapy gets the *Beauty* treatment. *Nature Biotech.* 20:987–988.

McDonald, J.F. 1993. Evolution and consequences of transposable elements. *Curr. Opin. Genet. Devel.* 3:855–864.

Galls and Goals in Plant Transformation

Birch, R.G. 1997. Plant transformation: problems and strategies for practical application. *Annu. Rev. Plant Physiol. Plant Mol. Biol.* 48:297–326.

Escobar, M., E.L. Civerolo, K.R. Summerfelt, and A.M. Dandekar. 2001. RNAi-mediated oncogene silencing confers resistance to crown gall tumorigenesis. *Proc. Natl. Acad. Sci. USA* 98:13437–13442.

Goodner, B. and others. 2001. Genome sequence of the plant pathogen and biotechnology agent *Agrobacterium tumefaciens* C58. *Science* 294:2323–2328.

Horsch, R.B., J.E. Fry, N.L. Hoffman, D. Eichholts, S.G. Rogers, and R.T. Fraley. 1985. A simple and general method for transferring genes into plants. *Science* 227:1229–1231.

Lurquin, P.F. 2001. *The Green Phoenix: A History of Genetically Modified Plants.* Columbia University Press, New York.

Sigee, D.C. 1993. *Bacterial Plant Pathology: Cell and Molecular Aspects.* Cambridge University Press, New York.

Tinland, B. 1996. The integration of T-DNA into plant genomes. *Trends Plant Sci.* 1:178–184.

Wood, D.W. and others. 2001. The genome of the natural genetic engineer *Agrobacterium tumefaciens* C58. *Science* 294:2317–2323.

Promoting Promoters and Constructing Constructs

Hayes, W. 1968. *The Genetics of Bacteria and Their Viruses.* Wiley, New York.

Russell, P.J. 2002. *Genetics.* Benjamin Cummings, New York.

Watson, J.D., M. Gilman, J. Witkowski, and M. Zoller. 1992. *Recombinant DNA.* Freeman, New York.

Reporter Genes

Berghammer, A.J., M. Klinger, and E.A. Wimmer. 1999. A universal marker for transgenic insects. *Nature* 402:370.

Chan, A.W.S., K.Y. Chong, C. Martinovich, C. Simerly, and G. Schatten. 2001. Transgenic monkeys produced by retroviral gene transfer into mature oocytes. *Science* 291:309–312.

Charng, Y.C., and A.J.P. Pfitzner. 1994. The firefly luciferase gene as a reporter for *in-vivo* detection of AC transposition in tomato plants. *Plant Sci.* 98:175–183.

Elliott, A.R., J.A. Campbell, B. Dugdale, R.I.S. Brettell, and C.P.L. Grof. 1999. Green fluorescent protein facilitates *in vivo* detection of genetically transformed plant cells. *Plant Cell Reports* 18:707–714.

Hare, P.D., and N-H. Chua. 2002. Excision of selectable marker genes from transgenic plants. *Nature Biotech.* 20:575–580.

Lewis, R. 1994. Refinements in bioluminescence assays expand technique's applications. *The Scientist* 8(5):17.

Wood, K.V., Y.A. Lam, H.H. Seliger, and W.D. McElroy. 1989. Complementary DNA coding click beetle luciferases can elicit bioluminescence of different colors. *Science* 244:700–702.

Test-tube Gene Cloning

Chien, A., D.B. Edgar, and J.M. Trela. 1976. Deoxyribonucleic acid polymerase from the extreme thermophile *Thermus aquaticus. J. Bacteriol.* 127:1550–1557.

Mullis, K. 1998. *Dancing Naked in the Mind Field.* Vintage Books, New York.

Mullis, K., F. Faloona, S. Scharf, R. Saiki, G. Horn, and H. Erlich. 1986. Specific enzymatic amplification of DNA *in vitro*: the polymerase chain reaction. *Cold Spring Harbor Symp. Quant. Biol.* LI:263–273.

Saiki, R.K. and others. 1985. Enzymatic amplification of β-globin genomic sequences and restriction site analysis for diagnosis of sickle cell anemia. *Science* 230:1350–1354.

Saiki, R.K. and others. 1988. Primer-directed enzymatic amplification of DNA with a thermostable DNA polymerase. *Science* 239:487–491.

Hybridizing DNA Molecules

Southern, E.M. 1975. Detection of specific sequences among DNA fragments separated by gel electrophoresis. *J. Mol. Biol.* 98:503–517.

Knockouts and Resuscitations

Bains, W. 1998. *Biotechnology from A to Z.* Oxford University Press, Oxford.

Dawkins, R. 1987. *The Blind Watchmaker.* W.W. Norton, New York.

Pearson, H. 2002. Surviving a knockout blow. *Nature* 415:8–9.

Index